Integers, Fractions and Arithmetic

A Guide for Teachers

MSRI Mathematical Circles Library

Integers, Fractions and Arithmetic

A Guide for Teachers

Judith D. Sally
Paul J. Sally, Jr.

MSRI
Mathematical Sciences Research Institute
Berkeley, California

AMS
AMERICAN MATHEMATICAL SOCIETY
Providence, Rhode Island

2010 *Mathematics Subject Classification.* Primary 11Axx, 97B50, 11–01.

For additional information and updates on this book, visit
www.ams.org/bookpages/mcl-10

Library of Congress Cataloging-in-Publication Data

Sally, Judith D.
  Integers, fractions and arithmetic : a guide for teachers / Judith D. Sally, Paul J. Sally, Jr.,
MSRI Mathematical Sciences Research Institute, Berkeley, California.
    pages cm. – (MSRI/mathematical circles library ; 10)
  Includes index.
  ISBN 978-0-8218-8798-1 (alk. paper)
  1. Arithmetic–Study and teaching (Elementary). I. Sally, Paul J., Jr., 1933- II. Title.
QA135.6.S234   2013
372.7′2–dc23
                                                                            2012036092

Contents

To The Reader

This book orginated as notes for a series of mathematical seminars for teachers and students preparing for teaching careers. Since the seminars are ongoing, we have retained the seminar format here. Individual readers can imagine themselves as participants in such a seminar, or just regard each seminar as a new chapter in the book.

Introduction

Our book, *Integers, Fractions and Arithmetic,* is a comprehensive and careful study of the fundamental topics of K - 8 arithmetic. This guide aims to help teachers understand the mathematical foundations of number theory in order to strengthen and enrich their mathematics classes. There are numerous activities that are suitable for teachers to bring into their classrooms.

The far-reaching point of view taken in the twelve seminars comprising this book is meant to enhance teachers' expertise. It is not meant for students learning the subject for the first time.

These seminars are designed for the professional development of teachers. They are also intended to be used in teacher preparation programs for undergraduates and graduates. They are appropriate for educators conducting enrichment programs such as Math Circles for Teachers, as well.

Conversations between the seminar leader and the other participants are an essential component of these seminars. Participants' input is important. There are many Seminar Exercises, Seminar Discussions and Seminar/Classroom Activities throughout this book. Comments on each of these are included as part of the main text.

Here is a brief outline of the topics and special features of each of the twelve seminars.

• Seminar 1, **Number Systems,** introduces the concept of a number system and its arithmetic operations. The natural numbers, the whole numbers and the integers, for example, are much more than just static collections of numbers. Each of these collections has a framework mandated by the properties of its operations of addition and multiplication.

• Seminar 2, **Divisibility and Order in the Integers,** introduces the idea of "dividing evenly" or divisibility, a basic and important topic in the theory of numbers. The properties of divisibility and its interactions with addition and multiplication are explained. The concept of divisor is defined. As an introductory illustration of the connection between divisibility and cryptography, we present a secret code game for the classroom. Order and its properties are discussed, as is the representation of integers by points on the number line.

• Seminar 3, **GCD's and The Division Algorithm,** examines the greatest common divisor (gcd) of a pair of integers. Several methods for calculating the greatest common divisor of two integers, including the Euclidean Algorithm, are explained and numerous examples are given. The division algorithm, or integer division with remainder, is introduced. It is a technique that is employed in every one of the succeeding seminars. The greatest common divisor of two integers a and b is shown to be a linear combination $sa + tb$. A feature of this seminar is an algorithm, the coefficient algorithm, that calculates the integers s and t.

• Seminar 4, **Prime Numbers and Factorization Into Primes,** begins with the definition of a prime number. Very quickly, using the properties of order, a simple useful tool for recognizing primes is presented, as is the Sieve of Eratosthenes, a method for finding all primes less than a fixed number. The sieve serves as an engaging activity for the middle grade classroom. Euclid's proof that there are infinitely many primes is discussed and the Fundamental Theorem of Arithmetic is explained. This theorem establishes the prime numbers as the "building blocks" of the integers.

• Seminar 5, **Applications of Prime Power Factorization,** explains how to use the prime power factorization of a positive integer $n > 1$, to count the number of divisors of n, and how to tabulate them. Of particular interest is the prime factorization of a perfect square. The information gleaned is used to study the "Locker Problem." The prime power factorization is used to calculate the greatest common divisor and the least common multiple of two positive integers, and the least common denominator of two positive fractions. The seminar concludes with more code games for students to show them how useful the understanding of primes and divisors can be.

• Seminar 6, **Modular Arithmetic With Applications to Divisibility Tests,** uses the examples of a light switch and a clock to introduce congruence and modular arithmetic. The concept of congruence is defined, and addition and multiplication of congruences (modular arithmetic) are explored. The properties of the operations of modular arithmetic are verified, and their consistency is discussed. The primary application of congruence in this seminar is the revelation that congruence is the basis of the popular classroom tests for divisibility.

• Seminar 7, **More Modular Arithmetic,** explores the notion of congruence classes for a particular modulus $m > 1$ and shows how the set of integers is partitioned into m nonintersecting congruence classes. The intriguing idea that addition and multiplication can be defined on the m congruence classes themselves leads to a number system, denoted $\mathbb{Z}_m$, with precisely m numbers. Linear congruences are defined and their solutions discussed.

• Seminar 8, **The Arithmetic of Fractions**, is the first of five seminars dedicated to fractions and decimals because of their importance in the classroom curriculum. The standard topics of addition, multiplication, division, common denominators and equivalent fractions are covered in detail in these five seminars. However, these topics are arranged in an order that is slightly different from the usual one. Multiplication is treated first, and with that in hand, common denominators and equivalent fractions are more readily understood and are available for use when discussing addition. Seminar 8 explains what a fraction is. The definition and properties of multiplication of fractions are developed. The concept of a common denominator of a pair of fractions, and two algorithms for finding the least common denominator are discussed. One uses the prime factoriztions of each of the two denominators. The other applies a formula involving only the product of the denominators and their greatest common divisor, allowing the calculation of the least common denominator by means of the Euclidean Algorithm. Equivalence of fractions is defined, its properties and many examples are studied. The cross product criterion for equivalence of two fractions is derived. The final topic in this seminar is equivalence classes of fractions and the fact that, in each equivalence class there is a fraction with smallest positive denominator. It is the fraction in "lowest terms."

• Seminar 9, **The Properties of Multiplication of Fractions**, begins by explaining the connection between a fraction in lowest terms and the more familiar concept of a fraction with numerator and denominator having no factors in common, other than ± 1. The important fact that multiplication respects equivalence is verified and the properties of multiplication are established. Other topics covered include mixed numbers and the division of fractions. The seminar end with a collection of "Word Problems." Four examples are given and three more are part of a Seminar/Classroom Activity. (Solutions are included as part of the text.)

• Seminar 10, **Addition of Fraction**, highlights the three step "Sure Fire Method" for finding the sum of two fractions. (The name derives from the fact that every problem on addition of fractions is solved the same way.) This method of addition is compared to the method using the least common denominator, and numerous examples are computed both ways. The properties of addition are verified. The additive inverse of a fraction is carefully explained and subtraction is defined in terms of the additive inverse. The essential distributive property that links multiplication and addition of fractions is discussed. The final topic establishes the consistency of addition of fractions with respect to equivalence.

• Seminar 11, **The Decimal Expansion of a Fraction,** defines the decimal expansion of a fraction a/b, with $0 < a < b$, as a sum of decimal fractions and describes how to construct the decimal expansion by means of repeated use of the division algorithm. (In Appendix A, the "long division algorithm" is derived from the same procedure.) Terminating and nonterminating decimals are defined. An important result is that a fraction a/b, in lowest terms, with $0 < a < b$, has a terminating decimal expansion precisely when its denominator b is a product of powers of 2 and/or 5. In the section on nonterminating decimals, the repetend of a/b, with $0 < a < b$, is defined as the sequence of place values of a/b that repeats. It is shown that the repetend begins at the 10^{-b}ths place or earlier. A highlight is the discussion of the fact that if b is relatively prime to 10, then the repetend begins in the tenths place, whereas if b is not relatively prime to 10 and has factors other than 2 or 5, then the repetend begins in the 10^{-j}ths place, where $j \geq 2$.

• Seminar 12, **Order and the Number Line,** defines order on the set of fractions as a natural extension of order on the integers, and verifies that equivalence respects order. The cross product comparison rule for fractions with positive denominators is derived. Comparison of fractions translates into a method for comparing decimals by comparing their place values. The facts that the set of positive fractions does not have a least element and that it is always possible to construct fractions between two given fractions are discussed. Next, attention is directed to the number line and the representation of fractions and decimals as points on this line. Order determines the position of these points on the number line. The distance between two nonequivalent fractions is defined, and it is seen that it is always positive. Furthermore, it is shown that the distance between two nonequivalent fractions can be estimated by comparing place values in their decimal expansions.

Acknowledgements. We acknowledge with gratitude the assistance of John Boller, Jonny Gleason, Nathan Hatch, Ryan Julian, Sara Mahoney, Xiao Xiao Peng, Weston Ungemach and Michael Wong. We thank the National Science Foundation for partial support, under Grant #NSF ESI-0101913, during the preparation of this manuscript.

Seminar 1

Number Systems

We set the conversational tone of these seminars by involving everyone in a mathematical discussion right from the start. We begin this first seminar with a Seminar/Classroom Activity. These activities are an integral part of each seminar and are suitable for use in the classroom.

✻

Seminar/Classroom Activity. For this activity, we focus on mathematical words related to numbers and arithmetic. The seminar leader will ask each of the participants, in turn, to write on the board a mathematical word associated with the topic Integers, Fractions and Arithmetic. Basic words, such as "sum" or "integer" are welcome. Complicated words are not so suitable here. Also, we suggest that the more variety of words put forward, the better and the more value for everyone. We hope mathematical conversations will take place as the words are put on the board.

After everyone has written a word on the board, the seminar leader will ask each participant to choose a word and use it in a sentence. For example, for the word "sum" we might say, "The sum of 3 and 5 is 8." For the word "integer" we might say, "-1 is a negative integer."

✻

Comments on the Seminar/Classroom Activity. (These comments are meant to be read after the activity is concluded.) Here are ten words associated to the topic Integers, Fractions and Arithmetic, and ten sentences using the words.

equation	negative
fraction	odd
exponent	subtract
less	factor
algorithm	prime

• The *equation* $2x = 3$ has the solution $x = 3/2$.
• A *negative* integer is represented by a point on the number line to the left of 0.

1

- The *fraction* $1/2$ is not an integer.
- 1 is an *odd* integer.
- The integer 7^8 has *exponent* 8.
- *Subtract* the amount of the withdrawl from the balance in your account.
- The integer -1 is *less* than the integer 1.
- 5 is a *factor* of every integer with units digit equal to 0 or to 5.
- An *algorithm* is a procedure for solving a problem in a finite number of steps.
- The integers 2 and 3 are *prime* numbers.

✳

These words, and many others, will be discussed in the course of these seminars. It is interesting to observe that many if not most of the words listed above refer not just to an individual number but to connections between numbers. For example, to tell whether an integer is prime or not, we must look at the integers that are its factors.

In this seminar, we explain the concept of number system, and examine familiar sets of numbers and their underlying structures. The natural numbers, whole numbers and integers, for example, are much more than just static collections of numbers. Each of these collections has a framework mandated by the properties of its operations of addition and multiplication.

1. Arithmetic in the Integers, Part I

The integers are far more than just the elements 0, 1, -1, 2, -2, ... The collection of integers, which we denote by $\mathbb{Z}$, is a set endowed with two operations, addition and multiplication, that have very important properties. Adding and multiplying are processes, but it is the rules of addition and multiplication that form the essence of arithmetic. We will define the word "operation" formally in Section 4 where we work with you to deduce some very important facts from the rules which we list now.

We restrict our attention to the integers in this section. First, we describe the rules for addition, then the rules for multiplication, and, finally, the essential rule, known as the distributive property, that connects addition and multiplication. The first rule formalizes the idea that when we add two integers, we obtain an integer.

(A1) Closure. If a and b are in $\mathbb{Z}$, then $a + b$ is in $\mathbb{Z}$.

For example, if we add the integers 53 and 101, the sum is the integer 154. If we add the integers 2 and -3, the sum is the integer -1.

If we want to add the integers 4, 6 and 21, we can first compute the sum $4 + 6 = 10$, and then add 21 to obtain $10 + 21 = 31$. Or, we can add 4 to the sum $6 + 21 = 27$ of 6 and 21 to obtain $4 + 27 = 31$. In both cases, the sum is the same. The associative rule, which we define next, tells us that this is always the case. It states that no matter how we associate, or group, numbers for addition, the sum is the same.

(A2) Associative Rule. If a, b and c are in $\mathbb{Z}$, then

$$(a + b) + c = a + (b + c).$$

Not all familiar operations on $\mathbb{Z}$ are associative. For an example of an operation that is *not* associative, consider exponentiation.

✳

Seminar Exercise. Find three integers a, b and c such that

$$(a^b)^c \neq a^{(b^c)}.$$

✳

Comments on the Seminar Exercise. One such example is $a = 2$, $b = 2$ and $c = 3$. In this case, $(2^2)^3 = 64$, whereas $2^{(2^3)} = 2^8 = 256$. Consequently, when you teach exponentiation, you need to be prepared to discuss associativity.

✳

The next rule states that we can add integers in any order.

(A3) Commutative Rule. For integers a and b,

$$a + b = b + a.$$

For example, $22 + 31 = 53$ and $31 + 22 = 53$. Also, $(-4) + 9 = 5$ and $9 + (-4) = 5$.

✳

Seminar Question. Is exponentiation commutative? Give an example that justifies your answer.

✳

Comments on the Seminar Question. Exponentiation is not commutative as the example $2^3 = 8$ and $3^2 = 9$ shows.

✳

Note that the combination of Rules (A2) and (A3) gives us many more options for grouping and ordering the sum of three numbers. For an example, we return to the sum of the three numbers 4, 6 and 21. Rules (A2) and (A3) allow us to first add 4 and 21 and then add 6. Here is one of several arguments showing why this is so.

$$\begin{aligned} (4 + 6) + 21 &= 4 + (6 + 21), \quad \text{by (A2)} \\ &= 4 + (21 + 6), \quad \text{by (A3)} \\ &= (4 + 21) + 6, \quad \text{by (A2)} \end{aligned}$$

This is a nice exercise for students who are learning the associative and commutative rules.

Be aware. The final two rules for addition of integers, (A4) and (A5), do not hold in some of the number systems we work with. We will discuss this in Section 3.

The next rule states that in the integers there is a particular integer with a very special property for addition. This integer is a solution x to the equation

$$a + x = a,$$

for all a in $\mathbb{Z}$. Such an element x is called an *additive identity*. The rule identifies the additive identity in $\mathbb{Z}$, and states that it is unique.

(A4) Additive Identity. There is precisely one element in $\mathbb{Z}$, called zero and denoted 0, such that

$$a + 0 = 0 + a = a, \text{ for all } a \text{ in } \mathbb{Z}.$$

Uniqueness means that there is only one solution, namely, the one we call 0, to the equation $a + x = a$, *for all a in* $\mathbb{Z}$. Consequently, if there is another integer z that satisfies $a + z = a$, for all integers a, then $z = 0$.

Since $\mathbb{Z}$ has an additive identity, 0, it makes sense to ask if, for each integer a, there is a solution x to the equation

$$a + x = 0.$$

The last rule for addition in the integers states that, for *every* integer a, there is precisely one such solution. The solution x depends on a, and is called the *additive inverse of a*.

(A5) Additive Inverse. For every a in $\mathbb{Z}$, there is a unique element, denoted $-a$, in $\mathbb{Z}$ such that

$$a + (-a) = 0.$$

Note that by (A3) we also have

$$(-a) + a = 0.$$

Uniqueness of the additive inverse of an integer a in $\mathbb{Z}$, means that if we find an integer b such that $a + b = 0$, then $b = -a$.

We put this idea to use immediately to find the additive inverse $-(-a)$ of the integer $-a$. We show that

$$-(-a) = a.$$

By definition, the additive inverse $-(-a)$ of $-a$, satisfies

$$-a + (-(-a)) = 0.$$

But by the definition of the additive inverse of a, w

$$-a + a = 0.$$

Thus, by uniqueness of the additive inverse, it follow

$$-(-a) = a.$$

Here are some examples.

(i) The additive inverse of 10 is the integer -10, sin

(ii) The additive inverse of 0 is 0. Since $0 + 0 = 0$, it

(iii) The additive inverse, $-(-7)$, of -7 is 7, since -7 -

We will formally define the notion of positive and nega
However, we want to point out now, that the examples al
additive inverse of an integer a can be positive or negati
the temptation to call $-a$ "negative a." As we have just s
inverse $-(-7)$ of the integer -7 is 7, which is not a negati
proper name for the additive inverse of the integer a is "mi $_{\smile\raisebox{-1pt}{$\scriptstyle\smile$}}$ u.

"Adding minus a to b" is precisely what we do when we subtract a from
b. We define the operation of *subtraction* in $\mathbb{Z}$ as follows. If a and b are
integers, then

$$b - a = b + (-a).$$

Thus subtraction is not a new operation. Subtracting a from b simply means
adding the additive inverse of a to b. For example,

$$5 - 2 = 5 + (-2) = 3.$$

Another example is

$$-5 - 2 = -5 + (-2) = -7.$$

But, be alert, the operation of subtraction has few of the properties that
addition has. For example, subtraction is neither associative:

$$(5 - 3) - 1 \neq 5 - (3 - 1)$$

nor commutative:

$$5 - 2 \neq 2 - 5.$$

Next, we investigate the rules for multiplication in the integers. The
first three properties are analogous to those for addition.

(M1) Closure. If a and b are in $\mathbb{Z}$, then $a \cdot b$ is in $\mathbb{Z}$.

For example, the product of $22 \cdot 39 = 858$, an integer.

(M2) Associative Rule. If a, b and c are in $\mathbb{Z}$, then

$$(a \cdot b) \cdot c = a \cdot (b \cdot c).$$

For example, $(2 \cdot 2) \cdot 3 = 12$ and $2 \cdot (2 \cdot 3) = 12$. Thus, multiplication is associative, whereas, we saw that exponentiation and subtraction are not.

(M3) Commutative Rule. For integers a and b,

$$a \cdot b = b \cdot a.$$

So $5 \cdot 3 = 15$ and $3 \cdot 5 = 15$. Also, $(-67) \cdot 4 = -268$ and $4 \cdot (-67) = -268$.

The integers have a *multiplicative identity*. In other words, the equation $a \cdot x = a$, for all integers a, can be solved for x. Moreover, there is a unique such solution.

(M4) Multiplicative Identity. There is precisely one element in $\mathbb{Z}$, denoted 1, such that

$$a \cdot 1 = 1 \cdot a = a, \text{ for all } a \text{ in } \mathbb{Z}.$$

✳

Seminar Exercise. $\mathbb{Z}$ has a multiplicative identity 1. Consequently, it makes sense to ask if there are integers that have a multiplictive inverse. What do you think is the definition of a multiplicative inverse for an element a of $\mathbb{Z}$? Find the integers, if any, that have a multiplicative inverse.

✳

Comments on the Seminar Exercise. If a is an integer, a *multiplicative inverse for a* is the unique *integer* x that satisfies the equation

$$ax = 1.$$

The integer 1 has a multiplicative inverse 1, since $1 \cdot 1 = 1$. The integer -1 has a multiplicative inverse -1, since $(-1) \cdot (-1) = 1$. Note that although $2 \cdot (1/2) = 1$, the integer 2 does not have a multiplicative inverse because $1/2$ is not an integer. In fact, 1 and -1 are the only integers that have a multiplicative inverse in the set of integers.

✳

Finally, we have the important rule that links addition and multiplication. Moreover, this is the *only* rule that links addition and multiplication. Its importance is hard to overstate.

(D) Distributive Rule. If a, b and c are integers, then

$$a \cdot (b + c) = (a \cdot b) + (a \cdot c);$$
$$(a + b) \cdot c = (a \cdot c) + (b \cdot c).$$

Observe that the second equation follows from the first by rule (M3). We look at two examples. The first is an example of left distributivity.

$$7 \cdot (-3 + 10) = (7 \cdot (-3)) + (7 \cdot 10) = (-21) + 70 = 49.$$

The next is an example of right distributivity.

$$(2 - 15) \cdot (-4) = (2 \cdot (-4)) + ((-15) \cdot (-4)) = (-8) + 60 = 52.$$

2. Roll Back, A Number Game of Chance

In this section we introduce a number game, called Roll Back, based on the rules of arithmetic. It has proved popular in the classroom.

Players: Roll Back can be played by groups of one to five students.

Supplies: One die, i.e., one of a pair of dice, for *each group* of players. A pencil and paper for *each player* in a group.

Object of the Game: The object of the game is to obtain a nonnegative balance by subtracting numbers from 100 based on rolls of the die.

Rules of the Game:

1. The players in each group take turns rolling the die which is rolled a total of five times. The players begin by writing the number 100 at the top of their papers.

2. After the first roll of the die, each player has a choice: either subtract from 100 the number rolled OR subtract from 100 the number rolled multiplied by 10. The result is called the first balance.

3. The most important rule of the game is that this balance and all subsequent balances must be nonnegative numbers.

4. After each of the four succeeding rolls of the die, each player has the same choice: either subtract from the previous balance the number rolled OR subtract from the previous balance the number rolled multiplied by 10. As each player makes these choices, the player must keep in mind that the balances must be nonnegative.

5. The student in each group with the nonnegative score closest to zero wins the game.

Here is an example.

Roll	Number Rolled	Choice	Balance
1	5	$5 \times 10 = 50$	100 - 50 =50
2	3	3	50 - 3 = 47
3	6	6	47 - 6 = 41
4	2	$2 \times 10 = 20$	41 - 20 = 21
5	3	3	21 - 3 = 18

After roll 3 and after roll 5, the player had no choice but to subtract the number rolled. After roll 2, the player could have chosen to subtract $3 \times 10 =$

30 from 50. Note that the players are doing some excellent mental, or pencil and paper, arithmetic as they make the choices required in the game.

✳

Seminar Exercise.

  (i) Assume you have the same rolls of the die as in the sample game. Can you roll back to a score closer to zero than in the sample game?

 (ii) Play another game of Roll Back.

✳

This game is one of many number games and classroom activities found in the authors' book *Trimathlon*. Games and classroom activities such as these play an important part of teaching arithmetic to elementary students.

3. Other Number Systems

The natural numbers,

$$\mathbb{N} = \{1, 2, 3, \ldots, n, \ldots\}$$

and the whole numbers,

$$\{0, 1, 2, 3, \ldots, n, \ldots\}$$

are subsets of the integers and inherit operations of addition and multiplication from $\mathbb{Z}$ because the sum and product of natural numbers are natural numbers, and the sum and product of whole numbers are whole numbers. Not all of the properties of these operations that hold for the integers also hold for the natural numbers and the whole numbers. We explore this fact in the next seminar exercise.

Note. If a set does not satisfy closure for a certain operation, then we do not discuss other properties for that operation on the set. If a set does not have an identity for a certain operation, then we do not discuss inverses for that operation.

✳

Seminar/Classroom Activity. For the following number systems, tell which of the rules of arithmetic (A1)–(A5), (M1)–(M4), and (D) hold and which do *not* hold. In the case of subsets of $\mathbb{Z}$, assume that addition and multiplication on the subset are the same as those in $\mathbb{Z}$. Justify your answers.

   (i) The natural numbers $\mathbb{N}$.
  (ii) The whole numbers.
 (iii) The set $\mathcal{E}$ of even numbers.
 (iv) The set $\mathcal{O}$ of odd numbers.
  (v) The set $\mathcal{S}$ of squares of whole numbers.
 (vi) This set is not a subset of $\mathbb{Z}$. Let $F = \{\mathcal{E}, \mathcal{O}\}$. Define two operations $+$ and $\cdot$ on F by means of the following operation tables.

+	$\mathcal{E}$	$\mathcal{O}$
$\mathcal{E}$	$\mathcal{E}$	$\mathcal{O}$
$\mathcal{O}$	$\mathcal{O}$	$\mathcal{E}$

$\cdot$	$\mathcal{E}$	$\mathcal{O}$
$\mathcal{E}$	$\mathcal{E}$	$\mathcal{E}$
$\mathcal{O}$	$\mathcal{E}$	$\mathcal{O}$

Read the table in the order *row* by *column*. For example, the sum of $\mathcal{E}$ and $\mathcal{O}$ is found at the intersection of row 2 and column 3, and the product of $\mathcal{O}$ and $\mathcal{E}$ is found at the intersection of row 3 and column 2.

✳

Comments on the Seminar/Classroom Activity.

(i) $\mathbb{N}$ does satisfy rules (A1)–(A3) and (M1)–(M4) and (D). $\mathbb{N}$ does not satisfy rule (A4). For there is no natural number x that satisfies the equation $1 + x = 1$. (We do not discuss additive inverses, since $\mathbb{N}$ does not have an additive identity.)

(ii) The whole numbers do satisfy (A1)–(A4) and (M1)–(M4) and (D). They do not satisfy rule (A5). There is no whole number x that satisfies the equation $1 + x = 0$.

(iii) The set $\mathcal{E}$ does satisfy (A1)–(A5) and (M1)–(M3) and (D). $\mathcal{E}$ does not satisfy rule (M4). For there is no even number x that satisfies the equation $2 \cdot x = 2$.

(iv) The set $\mathcal{O}$ does not satisfy rule (A1) because the sum of two odd numbers is even, so we do not discuss rules (A2)–(A5) or (D). The set $\mathcal{O}$ does satisfy rules (M1)–(M4.)

(v) The set $\mathcal{S}$ does not satisfy rule (A1) because the sum of two squares need not be a square. For example, $1^2 + 1^2 = 2$ is not a square. All multiplicative properties are satisfied.

(vi) The set F satisfies all of the rules (A1)–(A5), rules (M1)–(M4) and (D). The element $\mathcal{E}$ of F is the additive identity and the element $\mathcal{O}$ is the multiplicative identity.

✳

4. Arithmetic in the Integers, Part II

In Section 1, we described the rules for addition and multiplication in the integers. In this section, we will deduce some familiar arithmetical facts from the rules (A1)–(A5), (M1)–(M4) and (D). We begin with a formal definition of a binary operation on a set S. A *binary operation on S* or *operation on S*, for short, is a rule $\star$ that associates to every pair (hence the word "binary") of elements (a, b) in any set S another element, denoted $a \star b$, *in S*. Note that the definition of binary operation includes closure. As we have seen, addition and multiplication of pairs of integers are binary operations on $\mathbb{Z}$. Subtraction is a binary operation on $\mathbb{Z}$ but not on $\mathbb{N}$ or on the set of whole numbers, because neither $\mathbb{N}$ nor the set of whole numbers is closed under subtraction. The number system F introduced in the Seminar/Classroom Activity in Section 3 is very interesting because it has only two numbers $\mathcal{E}$ and $\mathcal{O}$. The addition and multiplication tables given in that activity define two binary operations on the number system F.

Note. From now on, we will use any of the notations ab, $(a)(b)$ or $a \cdot b$ to denote the product of the integers a and b. The choice is usually dictated by a wish for clarity.

When we solve certain equations involving addition, one of the procedures we might like to apply is cancellation. Should this be another rule, or does it follow from the ones we have?

✳

Seminar Exercise. Show how the property (AC), defined below, follows from rules (A1)–(A5).
(AC) Additive Cancellation. If

$$c + a = c + b, \quad \text{for} \quad \text{integers } a, \ b, \ \text{and } c,$$

then

$$a = b.$$

✳

Comments on the Seminar Exercise. We verify Property (AC) by adding the additive inverse, $-c$, of c to both sides of the given equation:

$$-c + (c + a) = -c + (c + b).$$

By additive associativity (A2), we have:

$$(-c + c) + a = (-c + c) + b.$$

By (A5), $-c + c = 0$, and the result is

$$0 + a = 0 + b.$$

By (A4), $0 + a = a$, and $0 + b = b$, so the equality $0 + a = 0 + b$ implies that

$$a = b.$$

Thus, the additive cancellation property, (AC), follows from the rules (A1)–(A5).

✳

Next, we look at multiplication by the additive identity 0.
Multiplication by 0. For all integers a,

$$a \cdot 0 = 0 \cdot a = 0.$$

Since 0 is the additive identity and the property we are to prove is one involving multiplication, we expect that the distributive property (D) which links addition and multiplication will play a role. First we apply (A4) in an unusual way: we "replace 0 by $0 + 0$". By (A4),

$$0 + 0 = 0.$$

Thus,

$$a \cdot (0 + 0) = a \cdot 0.$$

We apply (D) to the left side of the equation above to obtain

$$a \cdot (0 + 0) = (a \cdot 0) + (a \cdot 0).$$

It follows from the equation $a \cdot (0 + 0) = a \cdot 0$ that

$$(a \cdot 0) + (a \cdot 0) = a \cdot 0.$$

By (A4), we may add 0 to the right hand side above and retain the equality:

$$(a \cdot 0) + (a \cdot 0) = (a \cdot 0) + 0$$

Finally, we apply (AC), and we have the desired conclusion:

$$a \cdot 0 = 0.$$

✻

Seminar Question. Are there any pairs of integers (a, b) for which $a + b = a \cdot b$?

Challenge. If so, find all such pairs.

✻

Comments on the Seminar Question. Since $0 + 0 = 0 \cdot 0$, and $2 + 2 = 2 \cdot 2$, the answer to the question is "yes." The argument showing that these are the only such pairs is left for you to think about.

✻

Properties that involve an integer of the form "minus a," that is, $-a$, must be interpreted properly. Recall that $-a$ is defined to be the unique additive inverse of a. In other words, $-a$ is the integer with the property that

$$a + (-a) = 0 = (-a) + a,$$

and it is the only such integer. Consequently, to show that a certain integer b presented to us is equal to $-a$, for some integer a, what we must do is show that when we add a to b, the sum is zero. Here are three such examples.

(PAI) Properties of Additive Inverses in $\mathbb{Z}$.

(i) For all integers a,
$$(-1) \cdot a = -a.$$

(ii) For all integers a and b,
$$a(-b) = -(ab) \text{ and } (-a)b = -(ab).$$

(iii) For all integers a and b,
$$(-a)(-b) = ab.$$

In words, property (i) states that the integer $(-1) \cdot a$, the product of the additive inverse of the multiplicative identity and a, is the additive inverse of a. Consequently, we must show that when we add $(-1) \cdot a$ to a, we obtain the additive identity 0. Since $(-1) \cdot a$ is a product and we are asked to use it additively, we expect distributivity to come into play, and it does. First, we apply the rule (M4) for the multiplicative identity 1,

$$a + (-1) \cdot a = 1 \cdot a + (-1) \cdot a.$$

Next, we apply (D) to write

$$1 \cdot a + (-1) \cdot a = [1 + (-1)] \cdot a.$$

By (A5),

$$[1 + (-1)] \cdot a = 0 \cdot a.$$

We apply the multiplication by 0 property, to obtain

$$0 \cdot a = 0.$$

Thus, we may conclude that

$$a + (-1) \cdot a = 0,$$

and

$$(-1)a = -a.$$

In words, property (ii) states that the integer $a(-b)$ is the additive inverse of ab, and that the integer $(-a)b$ is the additive inverse of ab. To show that $a(-b)$ is the additive inverse of ab, we must show that $a(-b) + ab = 0$. We first apply distributivity, (D),

$$a(-b) + ab = a(-b + b).$$

By (A4),

$$-b + b = 0,$$

and, by the multiplication by 0 property,

$$a \cdot 0 = 0.$$

Thus, we may conclude that

$$a(-b) + ab = 0 \ \text{ and } a(-b) = -(ab)$$

A similar argument, which we leave to you, shows that $(-a)b = -(ab)$.

✳

Seminar Exercise. Show that

$$(-a)b = -(ab).$$

Then verify (iii): $(-a)(-b) = ab$.

✳

Comments on the Seminar Exercise. We must show that $(-a)b + (ab) = 0$. By (D),

$$(-a)b + ab = (-a + a)b.$$

By (A4),

$$-a + a = 0.$$

By the multiplication by 0 property,

$$0 \cdot b = 0.$$

Thus,

$$(-a)b + ab = 0 \ \text{ and } (-a)b = -(ab)$$

To prove (iii), we must show that $(-a)(-b) + -(ab) = 0$. But we have just shown that $-(ab) = (-a)b$, so, by the same reasoning as above,

$$(-a)(-b) + -(ab) = (-a)(-b) + (-a)b = (-a)[-b + b] = (-a)(0) = 0.$$

✳

Reference

Sally, J. and Sally, Jr., P., *Trimathlon*, A. K. Peters, Ltd., 2003.

Seminar 2

Divisibility and Order in the Integers

1. Introduction

We have discussed the operations of addition, multiplication and subtraction in the integers. Since, with the exception of 1 and -1, integers have no multiplicative inverses in $\mathbb{Z}$, there is no operation of division in $\mathbb{Z}$. However, for some pairs of integers, a rudimentary form of division exists and leads to fundamental results in the theory of numbers in $\mathbb{Z}$.

Observe that there are some pairs of integers, a and b, where a divides b "evenly," so that the quotient is an integer. Some examples of such pairs are 4 and 12, 5 and 125, 13 and 91, and 111 and 1998. The quotients are the integers 3, 25, 7 and 18, respectively.

Probably the most basic and important topic in the theory of arithmetic in $\mathbb{Z}$ is the idea of "dividing evenly." It is called *divisibility*. The precise definitions are as follows. If a and b are integers, we say that a *divides* b if there is an integer n such that

$$b = na.$$

The integer n is called the *quotient*. Occasionally, it is convenient to write the quotient n in the familiar fraction notation:

$$n = b/a.$$

The additive identity 0 divides 0, because $0 = n \cdot 0$, for every integer n. However, 0 does not divide any other integer, because $b = n \cdot 0$ implies that $b = 0$. We tacitly disregard this uninteresting case.

The examples above show that 4 divides 12, with $n = 3$; 5 divides 125, with $n = 25$; 13 divides 91, with $n = 7$; and 111 divides 1998 with $n = 18$. The integer 7 does not divide 169, because there is no integer n such that $7n = 169$, and the integer 8 does not divide 106 because there is no integer n such that $8n = 106$.

When the integers a and b are positive and a divides b, we can think of b as broken up into n groups of a elements each, or as broken up into a groups of n elements each. As elementary as this form of division is, it leads to fascinating information about integers.

When a divides b, we say that a *is a divisor or a factor of* b and b *is a multiple of* a. (The terms "divisor" and "factor" are interchangeable.) We write a divides b symbolically as follows:

$$a \mid b,$$

and a does not divide b as

$$a \nmid b.$$

Thus, $\mid$ means "divides" and $\nmid$ means "does not divide." The symbols are read from left to right as "a divides b" and "a does not divide b," respectively. Thus, $4 \mid 12$, but $5 \nmid 12$. Observe that the integers a and b may be positive or negative.

2. Properties of Divisibility

The following properties can be verified directly from the definition.

(R) Reflexivity For every integer a,

$$a \mid a.$$

To verify this, the definition requires that we find an integer n such that

$$a = na.$$

Since $a = 1 \cdot a$, we may take $n = 1$ and, it follows that $a \mid a$.

(D0) Every Integer Divides 0 For every integer a,

$$a \mid 0.$$

To check this property, we must show that $0 = n \cdot a$, for some integer n. Since $0 = 0 \cdot a$, we may take $n = 0$.

($\pm$1D) 1 and -1 Divide Every Integer For every integer a,

$$1 \mid a \quad \text{and} \quad -1 \mid a$$

(T) Transitivity If a, b and c are integers and if

$$a \mid b \text{ and } b \mid c, \text{ then } a \mid c.$$

(DAI) Divisibility and Additive Inverses For any integers a and b,
(i) $a \mid b$ precisely when $-a \mid b$, and
(ii) $a \mid b$ precisely when $a \mid (-b)$.
To verify a statement using the phrase "precisely when" such as "$a \mid b$ precisely when $-a \mid b$," means that we must demonstrate that the first statement, $a \mid b$, implies the second, $-a \mid b$, and that the second statement, $-a \mid b$, implies the first, $a \mid b$. Similarly, to verify "$a \mid b$ precisely when $a \mid (-b)$," we must verify that $a \mid b$ implies $a \mid (-b)$, and that $a \mid (-b)$ implies $a \mid b$.

✳

Seminar Exercise. Verify properties ($\pm$1D), (T) and (DAI).

※

Comments on the Seminar Exercise. To verify Property ($\pm$1D) for the integer 1, we need an equation of the form $a = n \cdot 1$, for some integer n. But by (M4), $a = a \cdot 1$, so we may take $n = a$. Similarly, to verify Property ($\pm$1D) for the integer -1, we have $(-a)(-1) = a$ by the Properties (PAI) in Seminar 1, Section 5. Thus, $-1 \mid a$, with $n = -a$.

Next, we show that transitivity, Property (T), holds. Since $a \mid b$, there is an integer k such that $b = ka$. Since $b \mid c$, there is an integer h with $c = hb$. Consequently, by associativity for multiplication, rule (M2),

$$c = hb = h(ka) = (hk)a.$$

Thus, we have $a \mid c$ with $n = hk$.

For Property (DAI), we verify the two statements required to demonstrate (i). Suppose that $a \mid b$. Then, there is an integer n such that $b = na$. Since $na = (-n)(-a)$, by the Properties (PAI) of additive inverses, it follows that $b = na = (-n)(-a)$, and that $-a \mid b$.

Suppose, on the other hand, that $-a \mid b$. If we apply what we have just verified to $-a$ instead of a, we have if $-a \mid b$, then $-(-a) \mid b$. But, by the Properties (PAI), $-(-a) = a$. We conclude that if $-a \mid b$, then $a \mid b$. The verification of (ii) is similar.

※

It follows from the above properties that every integer has at least one positive divisor, namely 1, and every positive integer other than 1 has at least two positive divisors, namely 1 and itself. The integer 12, for example, has a total of six positive divisors: 1, 2, 3, 4, 6 and 12. Note that the positive divisors of 12 come in pairs: 1 and 12, 2 and 6, and 3 and 4. If we return to the definition, we see this is always true. When a divides b, there is an integer n such that

$$b = na.$$

Since the quotient n is also a divisor of b, we can think of a and n as *codivisors* of b, or *cofactors* of b. If $a = n$, then

$$b = a^2,$$

and b is called a *perfect square*. It follows that if we find one divisor, a, of an integer b, we obtain a pair of divisors, a and its codivisor n. These codivisors are distinct except for one pair in the case where b is a perfect square. Here is a bit of poetry to help keep this in mind.

Divisors come in pairs, except for perfect squares.

※

Seminar/Classroom Activity. Find all the divisors of 144.

We will do this activity together. We proceed by "trial and error" to make a list of divisors and their codivisors. We know that 1 and 144 divide 144. How do we find other divisors of 144, if there are any? First, we note that 144 is even, so there are other divisors. In particular, 2 divides 144. We have

$$144 = 2 \cdot 72,$$

and 72, the codivisor of 2 also divides 144. Is 3 a divisor of 144? The answer is "yes," and we see that

$$144 = 3 \cdot 48,$$

so 48, the codivisor of 3 also divides 144. How about 4? We have

$$144 = 4 \cdot 36,$$

so 4 and 36 divide 144. Does 5 divide 144? No, because the units digit of every product $5 \cdot n$ is either 5 or 0, but the units digit of 144 is 4. The integer 6 divides 144, for

$$144 = 6 \cdot 24.$$

Does 7 divide 144? No, for we have that $7 \cdot 20 = 140$ and $7 \cdot 21 = 7 \cdot 20 + 7 = 147$. The integer 8 does divide 144, and we have

$$144 = 8 \cdot 18.$$

Next, we look at 9, 10 and 11. The positive integer 9 does divide 144, for

$$144 = 9 \cdot 16,$$

but 10 and 11 do not divide 144. The integer 10 does not because the units digit of 144 is not 0, and 11 does not because $11 \cdot 13 = 143$ and $11 \cdot 14 = 143 + 11 = 154$.
The integer 12 divides 144 and we have

$$144 = 12 \cdot 12 = 12^2.$$

The integer 144 is a perfect square.
We look at 13, next. Since $13 > 12$, if 13 divides 144, then its codivisor n would have to be less than 12, otherwise the product $13 \cdot n$ of 13 and n would be greater than $12^2 = 144$. But, we have already found all of the divisors of 144 that are less than 12, and none has codivisor 13. Thus, 13 does not divide 144. Once we reach a divisor of 144 that is greater than 12, it has a codivisor less than 12, so we can stop. We have the full list of all of the positive divisors of 144. There are 15 positive divisors. (We do not count 12 twice.) Since 144 is a perfect square, it has an odd number of positive

divisors.

1	144
2	72
3	48
4	36
6	24
8	18
9	16
12	12

Note that one integer in each pair of codivisors is less than or equal to 12, and the other is greater than or equal to 12.

✳

In the classroom, divisors, greater than 1, and their codivisors can be represented pictorially by means of factor trees. Here are some examples.

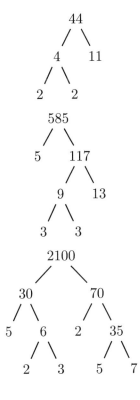

Note that factor trees for the same number may have many different branches, but the collection of leaves at the end of each branch is the same. The leaves are prime numbers. Prime numbers are positive integers greater

than 1 that have no positive divisors, i.e., factors, other than 1 and themselves. Prime numbers are the topic of Seminars 4 and 5.

✳

Seminar/Classroom Activity Draw factor trees for the following integers:

$$30, \quad 143, \quad 48$$

✳

Comments on the Seminar/Classroom Activity.

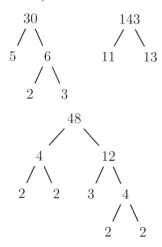

3. Divisibility and Arithmetic in $\mathbb{Z}$.

Arithmetic in $\mathbb{Z}$ is based on the operations of addition and multiplication.
How does divisibility interact with these operations? For example, 7 divides
84 and 7 divides 56, and, in fact 7 divides the sum $84 + 56 = 140$. The integer
8 divides 96. It also divides $96 \cdot 5 = 480$. These examples are not anomalies,
as we see now. To verify the general statements, apply the definition of
divisibility and the rules for addition and multiplication.

✳

Seminar Exercise.
  (i) Suppose that a, b and c are integers such that $a \mid b$ and $a \mid c$. Show
      that $a \mid (b + c)$.
 (ii) Suppose that a, b and c are integers such that $a \mid b$. Show that $a \mid bc$.
(iii) Let a, b and c be integers such that $a \mid b$ and $a \mid c$. Show that $a \mid$
      $(hb + kc)$, for all integers h and k.

✳

Comments on the Seminar Exercise. For (i), we have $b = ma$ and
$c = na$, for some integers m and n. Consequently,

$$b + c = ma + na = (m + n)a.$$

By closure $m + n$ is in $\mathbb{Z}$, so $a \mid (b + c)$.

For (ii), we have $b = ma$, for some integer m. By (M2) and (M3),

$$bc = (ma)c = (mc)a.$$

Since, by (M1), mc is in $\mathbb{Z}$, we have $a \mid bc$. In particular, we see that $a \mid -b$ because $-b = (-1)b$.

Properties (i) and (ii) can be combined to show (iii). It follows from (ii) that we have $a \mid hb$ and $a \mid kc$, and it follows from (i) that $a \mid (hb + kc)$.

✳

An expression such as $hb + kc$ is called a *linear combination of b and c*. For example, 13 is a linear combination of 5 and 3, since $13 = 2 \cdot 5 + 1 \cdot 3$, and -1 is a linear combination of 5 and 3, since $-1 = (-2) \cdot 5 + 3 \cdot 3$.

4. An Activity on Secret Codes, I

One of the most interesting applications of number theory is in the field of cryptography, the study of secret communication. Many useful codes are based on the idea that, if an integer is very large, it is very difficult to find its divisors. The following classroom game serves as an introductory illustration of the connection between codes and divisors. We will return to this game again in the course of these seminars.

We set up a correspondence between the letters of the alphabet and the integers from 1 to 26 as follows.

A	B	C	D	E	F	G	H	I	J	K	L	M	N	O	P	Q
1	2	3	4	5	6	7	8	9	10	11	12	13	14	15	16	17

R	S	T	U	V	W	X	Y	Z
18	19	20	21	22	23	24	25	26

Using the table above, we assign a number to a word (in English) this way: the number is the product of the numbers assigned to each of the letters of the word. The number corresponding to the word HELP, for instance, is the product of H = 8 and E = 5 and L = 12 and P = 16 which is 7680. To put any word into code, or to "encode" the word, simply multiply the numbers that correspond to the letters of the word together to form the code number. Consequently, these numbers are divisors of the code number. In the example above, the code number for HELP is 7680, and 8, 5, 12 and 16 are divisors of 7680. Code numbers assigned to words become large very quickly. For instance, the three letter word ZIP has code number 3744.

✳

Seminar Code Game 1. Take out your calculators. What is the largest code number you can find that is assigned to a three letter (English) word? What is the largest code number you can find that is assigned to a four letter (English) word?

✳

A more challenging way to play games with this code is to have the players use the table above to "decode" a given number into a corresponding (English) word. For example, the code number 40, corresponds to the word BAT and several other words. How do we see that 40 corresponds to BAT? The letters of the code word correspond to positive divisors of 40 that are less than 27. The positive divisors of 40 less than 27 are 1, 2, 4, 5, 8, and 20. The corresponding letters are: A, B, D, E, H, and T.

To form code words with code number 40, we must combine these letters in various ways so that they form a word and so that the product of numbers corresponding to the letters of the word is 40. The word HAT uses these letters, but its code number is 160, way too big. The word BED has code number 40. So far we have BAT and BED, both with code number 40. You see that secret messages with our simple code can be confused and distorted. This is due to the fact that there many ways to write a number as a product of factors.

Since the letter A corresponds to the multiplicative identity and the operation we are using is multiplication, you are free to use as many A's in your word as you please. Can you find another word with code number 40?

When we play this more challenging game with the code, it is sensible, for the first several rounds, to use numbers with two or three digits. We will play the game with secret messages later, after we have delved deeper into the topic of divisors.

✳

Seminar Code Game 2. Each player needs a pencil and a piece of paper. One person is chosen to be the transmitter of the code.
Object of the Game: The object of the game is to use the code given in the table above to decode the number given by the transmitter into a one word message. The winner is the first player to accomplish this. (Some appropriate codes for this first round of the game are: 16, 100, 625, 15, 245, 35, 36.)

✳

Let's look at the number 385. It obviously has 5 as a divisor: $385 = 5 \cdot 77$ and $77 = 7 \cdot 11$. The numbers 5, 7 and 11 are special numbers. Each is what is called a *prime number* because it has only 1 and itself as positive divisors. The other divisors of 385 are products of 5, 7 and 11. Consequently, if 385 is a code number, it has very few divisors that are less than 27. The only possible letters in a corresponding code word are A, E, G and K. It does not take long to come up with a code word KEG. This illustrates that prime numbers are very useful in games based on our code. In fact they are the building blocks of the theory of numbers in $\mathbb{Z}$.

5. Order in $\mathbb{Z}$

The fact that we can compare any two integers with respect to order is one of the most beautiful properties of $\mathbb{Z}$. To explain how order is defined, we formally define the word positive. An integer is *positive* precisely when it is a natural number. Consequently, *the set of positive integers* is the set

$$\{1, 2, 3, \ldots, n, \ldots\}.$$

Thus, the set of natural numbers and the set of positive integers are one and the same and the set of positive integers is closed under addition and multiplication. See Seminar 1, Section 4.

    If a and b are integers, we say that a *is less than* b, in symbols $a < b$, if there is a positive integer c such that

$$a + c = b.$$

Note that $c = b - a$, so an alternate form of the definition is: $a < b$ if $b - a$ is a positive integer, that is, a natural number.

    For example, $5 < 8$, because $5 + 3 = 8$. In this example, $c = 3$. Using the alternate definition, we have that $8 - 5 = 3$ is a positive integer, so $5 < 8$. For a more interesting example, we see that $-6 < -2$ because $-6 + 4 = -2$. Here the c in the definition is 4. Using the alternate definition, we have $-6 < -2$ because $-2 - (-6) = -2 + 6 = 4$ is a positive integer. (Observe that we applied the properties of additive inverses (PAI) derived in Seminar 1, to obtain $-(-6) = 6$.)

    We define "greater than" in terms of "less than." If a and b are integers, we say that b *is greater than* a, in symbols $b > a$, if a is less than b, i.e., if $a < b$. So, $b > a$ means exactly the same thing as $a < b$, and thus, $b > a$ precisely when $b - a$ is a positive integer.

    Note that, by definition, $b > 0$ means that the integer $b - 0$ is positive. But, by (A4), $b - 0 = b$, so $b > 0$ is true precisely when b is positive. Consequently, the following property holds.

Positivity Property. An integer b is positive precisely when $b > 0$.

✳

Seminar Exercise. Verify the following three properties of order.
(O1) Transitivity. Let a, b and c be integers. If $a < b$ and $b < c$, then $a < c$.
(O2) Additivity Property. If a, b and c are integers, and $a < b$, then $a + c < b + c$.
(O3) Scaling by a Positive Integer Property. If a, b and c are integers with $a < b$ and $c > 0$, then $a \cdot c < b \cdot c$.

✳

Comments on the Seminar Exercise. We apply the definitions.
For (O1), the hypotheses $a < b$ and $b < c$ imply that $b - a > 0$ and $c - b > 0$.
To show that $a < c$, we need to see that $c - a > 0$. But, the sum of two
positive integers is positive. So, $b - a + (c - b) = c - a > 0$.
For the additive property of order (O2), we must show that $b + c - (a + c) > 0$.
Using the rules of arithmetic, we have

$$b + c - (a + c) = b + c - a + (-c).$$

So,

$$b + c - (a + c) = b - a + c + (-c).$$

By (A5), $c + (-c) = 0$. It follows, by (A4), that

$$b + c - (a + c) = b - a + 0 = b - a.$$

However, $b > a$, so $b - a > 0$ and

$$b + c - (a + c) > 0,$$

exactly what we needed to show.
Observe that Property (O3) states that when we scale integers by a positive
factor, we retain their relative sizes. This property follows from the multiplicative closure of the set of positive integers. For if $a < b$ and $c > 0$, then
$b - a$ and c are positive integers. Consequently, $c(b - a)$ is a positive integer,
and, by distributivity, $cb - ca > 0$. Thus, $ca < cb$.

✳

What are the negative integers? An integer a is *negative* if $a < 0$.
Negativity Property. An integer a is negative if and only if its additive
inverse $-a$ is positive.

To see why this property is true, note that if $a < 0$, then $0 - a = -a > 0$,
so $-a$ is positive. Conversely, suppose $-a > 0$. To show that $a < 0$, we need
to show that $0 - a > 0$, but this is precisely our hypothesis, so $a < 0$. For
example, the integer -4 is negative. Its inverse, $-(-4) = 4$ is positive.

We use the negativity property to identify the set of negative integers.
If $a < 0$, then $-a > 0$, so $-a = n$, for some natural number n. It follows
that $a = -(-a) = -n$, and that the set of negative integers is the set

$$\{-1, -2, -3, \cdots - n, \ldots, \}.$$

The next property shows how order impacts additive inverses.
(O4) Order and Additive Inverses. For any two integers a and b, if

$$\text{if } a < b, \text{ then } -b < -a.$$

Moreover, the converse is also true:

$$\text{if } -b < -a, \text{ then } a < b.$$

The reason is this. We have $a < b$. By (O2), we may add $-a - b$ to both sides:

$$a - a - b < b - a - b.$$

Thus, by (A2) and (A5), $-b < -a$.
For the converse, if $-b < -a$, then, by (O2), we may add $a + b$ to both sides:

$$a + b - b < -a + a + b.$$

Thus, by (A2) and (A5), $a < b$.

Observe that, by taking $a = 0$, Property (O4) subsumes the negativity property.

The relation of order on the integers has two additional very important properties. The first is known as the trichotomy property.
(T) Trichotomy. For integers a and b, exactly one of the following holds: $a < b$, $a = b$, $a > b$.
Since $a < b$ means $b - a > 0$ and $a > b$ means $a - b > 0$, the trichotomy property can be restated in the following way.
(T) Trichotomy (alternate form). For any integer a, one and only one of the following holds: $a > 0$, $a = 0$ or $a < 0$, i.e., exactly one of the following holds: a is positive, a is zero or a is negative.

Note. We write $a \leq b$ as shorthand for a is less than b or a is equal to b. It follows immediately from the trichotomy property that if $a \leq b$ and $a \geq b$, then $a = b$.

The property of trichotomy signifies that every pair of integers can be compared with respect to order and implies that given a basket full of integers, we can dump them out and line them up in order. Consequently, the set of integers is said to be *totally ordered*.

The fact that $\mathbb{Z}$ is totally ordered gives rise to the powerful graphic representation of the integers as points on the number line. In fact, every point on this line can be interpreted as a real number, but, for now, we focus on the points representing integers.
It is the properties of order in the integers that dictate how we mark the points on this line. We begin by drawing a horizontal line and marking a point labeled 0 on it. This point corresponds to the integer 0. To the right of 0, we mark off another point which we label 1. This point corresponds to the integer 1. (The placement of 1 to the right of 0 is a convention.) The distance between 0 and 1 is our unit of measurement. To represent integers on this line, we mark off points to the right of 1 and to the left of 0, with the distance between adjacent points equal to one unit. The points to the right of 0 correspond, from left to right, to the positive integers *in ascending order*, and are labeled as such. Points to the left of 0 correspond, from right

to left, to the negative integers *in descending order* and are labeled as such. Note that it is the fact that the integers are *totally ordered* that permits this representation on the number line. Frequently, the points on the number line that represent integers are called *integer points*.

Note how isolated the integer points are from one another. Each point between an integer point and its adjacent integer point represents a real number, but none are integers and we will not think about them now except to know they are there.

The second property for order in $\mathbb{Z}$ is crucially important for all of mathematics. It is called the "Well-Ordering" property.

(WO) The Well Ordering Property. In every nonempty subset of the *positive* integers there is a smallest positive integer.

Let us look at some examples.

(i) The set of all positive integers, the natural numbers $\mathbb{N}$, has smallest positive integer 1.

(ii) The set $2\mathbb{N}$ of positive multiples of 2 :

$$2\mathbb{N} = \{2, 4, 6, 8, \ldots, 2n, \ldots\}$$

is a subset of $\mathbb{N}$. It is clear that 2 is the smallest element of the set $2\mathbb{N}$. The subset

$$\{n \text{ in } \mathbb{Z} \mid n > 3.99999\}.$$

of $\mathbb{N}$ has smallest element 4.

The fact that the integer points on the number line seem isolated from one another is due to the Well Ordering Property which is used to prove that there is no integer between 0 and 1. We can mark off a small interval around each integer point in which there are no other integer points. The representation of fractions on the number line, which we will discuss in Seminar 12, reflects much different behavior.

Many of the sets we study do not have the Well Ordering Property. For example, as the integer points on the number line illustrate so well, $\mathbb{Z}$ does not have the well ordering property. There in no smallest integer in $\mathbb{Z}$. As we will see in Seminar 12, the set of positive fractions does not have the Well Ordering Property.

✳

Seminar Exercise. Give an example of a nonempty set that has the Well Ordering Property, and an example of a set that does not have this property. Explain your answers.

✳

To conclude this section, we look at two statements regarding order and negative numbers that are rather delicate. Remember that an integer c is negative if and only if its additive inverse $-c$ is positive, i.e., satisfies $-c > 0$.

(O5) Scaling by a Negative Integer Property. If a, b and c are integers with $a < b$ and $c < 0$, then $a \cdot c > b \cdot c$.

Scaling by a negative integer is a counterintuitive notion. Nevertheless, Property (O5) follows from our positive scaling property (O3). Since $c < 0$, it follows that $-c > 0$. Consequently, by Property (O3), $-c \cdot a < -c \cdot b$. But, by applying the properties of additive inverses (PAI), verified in Seminar 1, Section 5, we have that

$$-c \cdot a = -(ca) \text{ and } -c \cdot b = -(cb).$$

Thus, $-(ca) < -(cb)$, and, by Property (O4), it follows that $ca > cb$.

(O6) The Product of Two Negative Integers. The product of two negative integers is a positive integer.

For the verification of Property (O6), we take two negative integers a and b. Their additive inverses are positive: $-a > 0$ and $-b > 0$. Since the positive integers are closed under multiplication, we have $(-a)(-b) > 0$. By (PAI), $(-a)(-b) = ab$. Hence, $(-a)(-b) = ab > 0$.

To illustrate the use of trichotomy in the demonstration of statements about the integers, we show that, under one very important condition, multiplicative cancellation is valid in the integers. Recall from Seminar 1, Section 5, that (AC), additive cancellation follows directly from the basic rules (A1)–(A5).

(MC) Multiplicative Cancellation Let a, b and c be integers with $a \neq 0$.

$$\text{If } ab = ac, \text{ then } b = c.$$

Do you see that the conclusion may be false if $a = 0$?

To justify Property (MC), we will show that if the conclusion $b = c$ is false then the hypothesis $ab = ac$ is also false. Thus it will follow that the conclusion is true and b and c must be equal.

Suppose that $b \neq c$. Since $a \neq 0$, it follows by trichotomy that $a > 0$ or $a < 0$. Again by trichotomy, since $b \neq c$, it follows that $b < c$ or $b > c$. We are in a position to apply our scaling results. There are several cases to examine. We will show that in no case is $ab = ac$. Suppose first $a > 0$. Then, our result on scaling by a positive integer gives $ab < ac$ if $b < c$, and $ab > ac$ if $b > c$. Suppose next that $a < 0$. Then, our result on scaling by a negative integer gives $ab > ac$ if $b < c$, and $ab < ac$ if $b > c$. Thus, $ab \neq ac$. Consequently, if $ab = ac$, the conclusion $b = c$ must be true.

Observe that in order to prove multiplicative cancellation in $\mathbb{Z}$, we need $\mathbb{Z}$ to be totally ordered.

✳

Seminar Exercise. Show that if a and b are integers and $ab = 0$, then $a = 0$ or $b = 0$.

✳

Comments on the Seminar Exercise. We are given

$$ab = 0.$$

If $a = 0$, our work is done. So we assume that $a \neq 0$. In Seminar 1, Section 5, we verified that $a \cdot 0 = 0$. Consequently, we have

$$a \cdot b = 0 = a \cdot 0.$$

Since $a \neq 0$, by property (MC), a may be cancelled and we have $b = 0$.

✳

Another very useful way to phrase the property we have just verified follows.

(NZ) Product of Nonzero Integers Property If a and b are integers with $a \neq 0$ and $b \neq 0$, then $ab \neq 0$.

✳

Seminar Exercise. Show that if a is *any* nonzero integer, then $a^2 > 0$.

✳

Comments on the Seminar Exercise. By trichotomy, we know that either $a > 0$ or $a < 0$. If $a > 0$, then scaling by a gives $a \cdot a = a^2 > 0$. If, on the other hand, $a < 0$, then $-a > 0$. So, by (PAI) and the first case, $(-a)(-a) = a^2 > 0$.

✳

6. Divisibility and Order in $\mathbb{Z}$.

We intuit that a divisor of a positive integer n cannot be greater than n, but now we are able to justify this property as well as other divisibility properties which require an understanding of order.

(CDD) Compare Size of Divisor and Dividend Property. If a and b are positive integers and if $a \mid b$, then $a \leq b$.

We verify (CDD). Since $a \mid b$, there is a positive integer n such that $b = na$. Since n is a positive integer, $n \geq 1$. If we multiply both sides of this inequality by the positive integer a, we have, by positive scaling, $na \geq a$. Thus, $b \geq a$.

Note. It follows from this result that

   (i) if a is a nonnegative integer and b is a positive integer greater than a
       such that $b \mid a$, then $a = 0$. (Why?)
  (ii) if the positive integer a divides 1, then $a = 1$. (Why?)

✳

Seminar Exercise. Suppose two positive integers a and b satisfy $a \mid b$ and
$b \mid a$. Show that $a = b$.

✳

Comments on the Seminar Exercise. Here is one way to do this exercise. By (CDD), $a \mid b$ implies $a \leq b$, and $b \mid a$, implies $b \leq a$. By trichotomy, $a \leq b$ and $b \leq a$ implies that $a = b$. (In more detail, if $a \neq b$, then we have $a > b$ and $b > a$, in contradiction to trichotomy.)

Seminar 3

GCD's and The Division Algorithm

1. Introduction

Divisibility gives us information about the connections between two integers, a and b, even in the case when neither one divides the other. Since 1 divides both a and b, we can ask what is the largest integer that divides both a and b. Consider the integers 12 and 54, for example. Neither one divides the other, but both are divisible by 1, 2, 3 and 6. The largest divisor common to 12 and 54 is 6.

✳

Seminar Questions What is the largest divisor common to the integers 10 and 25? What is the largest divisor common to the integers 32 and 36? (The answers are found in the text below.)

✳

The positive integer d is the *greatest common divisor of the positive integers a and b* if $d \mid a$ and $d \mid b$, and if every positive integer dividing both a and b is less than or equal to d. The greatest common divisor of a and b, is denoted by $d = \gcd(a, b)$. (Note that it follows from the definition that $\gcd(a, b) = \gcd(b, a)$.) We have seen that the greatest common divisor of 12 and 54 is 6. We write $\gcd(12, 54) = 6$. We also have $\gcd(10, 25) = 5$, $\gcd(32, 36) = 4$ and $\gcd(3, 4) = 1$.

The case where $\gcd(a, b) = 1$, for a pair of positive integers a and b, is especially interesting. It means that a and b have no divisors in common other than 1. We say that positive integers a and b are *relatively prime* if $\gcd(a, b) = 1$. The following pairs of integers are relatively prime: 3 and 4, 25 and 36, and 7 and 39. The integer 7 has no divisors other than itself and 1, so 7 and 39 are relatively prime. For that matter, so are 7 and any positive integer that is not a multiple of 7. We stress that the notion of "relatively prime" is a property of *pairs* of integers. (It does not make sense to say that a single integer is relatively prime. The pair of integers 9 and 10 are relatively prime, but the pair 9 and 21 are not.)

Recall that, in these seminars, the words *divisor* and *factor* are interchangeable. The greatest common divisor is also known as the greatest common factor.

We will develop techniques for computing the greatest common divisor. For now, to find $\gcd(a, b)$, find the divisors of, for example, the smaller of the two positive integers, say b, and check which, if any, are divisors of a. Consider, for example, the positive integers 30 and 75. The numbers 1, 2, 3, and 5 are divisors of 30, and all other divisors of 30 are products of these divisors. The integers 1, 3 and 5 are also divisors of 75, so the product 15 is a divisor common to both 30 and 75. Since 2 is not a divisor of 75, it is clear that $\gcd(30, 75) = 15$. For another example, consider the integers 5 and 21. The divisors of 5 are 1 and 5. Since $5 \nmid 21$, it follows that $\gcd(5, 21) = 1$.

✳

Seminar Exercise. Compute $\gcd(48, 66)$ and $\gcd(17, 153)$.

✳

Comments on the Seminar Exercise. Since $66 = 6 \cdot 11$, and $48 = 6 \cdot 8$, we see that $\gcd(48, 66) = 6$. Since $153 = 17 \cdot 9$, the integer 17 divides 153, so it follows that $\gcd(17, 153) = 17$.

✳

Recall that a linear combination of a and b is an integer of the form $ha + kb$, where h and k are integers. A linear combination of positive integers is an integer, but it may be positive, negative or zero. In our discussion of the gcd we will use the fact, found in Part (iii) of the Seminar Exercise in Section 3 of Seminar 2, that if the integer c is a common divisor of integers a and b, then c divides every linear combination of a and b. In other words, if $c \mid a$ and $c \mid b$, then, $c \mid (ha + kb)$, for any integers h and k.

The following observation is very useful. Remember that if $d = \gcd(a, b)$, then $d \mid a$ and $d \mid b$, so the quotients a/d and b/d are integers.

OBSERVATION 1.1. *Let a and b be positive integers, and set $d = \gcd(a, b)$. Then*

$$\gcd\left(\frac{a}{d}, \frac{b}{d}\right) = 1.$$

To verify Observation 1.1, we show that 1 is the largest integer that divides a/d and b/d. Since $d = \gcd(a, b)$, the integer d is the largest positive integer that divides a and b. Let c be a positive integer and suppose that c divides a/d and b/d. We aim to show that $c = 1$. We have $a/d = rc$ and $b/d = sc$, for some positive integers r and s. Consequently, we have equations

$$a = rcd \quad \text{and} \quad b = scd$$

which show that cd divides both a and b. Thus, $cd \leq d$. But d divides cd, and, by Property (CDD) in Seminar 2, Section 6, it follows that $d \leq cd$. Thus, $cd \leq d$ and $d \leq cd$, so by tricotomy, $cd = d$. By Property (MC), it follows that $c = 1$.

Here is another useful observation about the gcd.

OBSERVATION 1.2. *Let a, b and d be positive integers. If d is a linear combination of a and b and if d divides both a and b, then $d = \gcd(a, b)$.*

We verify the observation as follows. The positive integer d can be written

$$d = ha + kb,$$

for some integers h and k. By hypothesis, $d \mid a$ and $d \mid b$, so it remains to show that d is the greatest integer that divides both a and b. If c is another common divisor of a and b, then, as we have recalled above, $c \mid d = ha + kb$. By Property (CDD) in Seminar 2, Section 6, $c \leq d$, which is the desired conclusion.

We will see later that $\gcd(a, b)$ is always a linear combination of a and b.

Presently, we will discuss an algorithm for computing the greatest common divisor of any two positive integers. It is named in honor of the Greek mathematician Euclid who lived around 300 BC. The algorithm was devised, most probably, well before Euclid's time. To prepare for the Euclidean Algorithm, we introduce a game about greatest common divisors, called Euclid's Game, which is essentially the same as that found on the website www.cut-the-knot.org.

2. Euclid's Game

The game is for two players. (Of course, one of the players can be you, the teacher, and the other, the class.) We illustrate the game with the pair of positive integers 280 and 105. To start, the numbers 280 and 105 are written on the board (or on a piece of paper.)

The First Play. The first player, Player A, subtracts *the smaller number* 105 *from the larger* 280, and writes the positive difference $280 - 105 = 175$ on the board.
The Second Play. The second player, Player B, selects any two of the three numbers now on the board and takes the positive difference *by subtracting the smaller number from the larger*. There are two possibilities.
(i) If the number is not already written on the board, Player B writes that difference down.
(ii) If the number is already on the board, Player B takes the positive difference of another pair of numbers written on the board.
Suppose player B chooses to take the difference of 175 and 105 to obtain $175 - 105 = 70$. Now there are four numbers on the board: 280, 105, 175

and 70.

Player A	Player B	Numbers on the Board
$280 - 105 = 175$		280, 105, 175
	$175 - 105 = 70$	280, 105, 175, 70

Continuation of Play. The players continue alternating turns taking positive successive differences, *by subtracting the smaller from the larger,* of any two of the numbers on the board until it cannot be done anymore because every such difference already appears. The game is finished.

Before we say who wins the game, we explain the signifigance of the game.

The Significance of Euclid's Game.

The smallest number on the board when the game is finished, which in this case is 35, is gcd(280, 105).

Observe that every positive successive differences taken in the game is a linear combination of 280 and 105, and thus, gcd(280, 105) divides the smallest number on the board. We will show that gcd(280, 105) is equal to the smallest number on the board in Section 4. In the meantime, we regard it as fact, and continue to play the game.

The winner of Euclid's Game is the first player whose difference produces the smallest number on the board. Here's an illustration of a possible continuation and finish of the game begun above.

Player A	Player B	Numbers on the Board
$280 - 105 = 175$		280, 105, 175
	$175 - 105 = 70$	280, 105, 175, 70
$105 - 70 = 35$		280, 105, 175, 70, 35
	$175 - 35 = 140$	280, 140, 105, 175, 70, 35
$280 - 35 = 245$		280, 245, 175, 140, 105, 70, 35
	$245 - 35 = 210$	280, 245, 210, 175, 140, 105, 70, 35

With eight numbers on the board, the game ends because the difference of any pair of these numbers yields another one of the numbers. Player A found the smallest difference, gcd(280, 105) = 35, first, so Player A wins the game.

※

Seminar Activity. Play Euclid's Game with the following pairs of integers: 90 and 42, and 84 and 70. Identify gcd(90, 42) and gcd(84, 70). The smallest number on the board when the game is finished is the greatest common divisor.

✳

To justify our claim that Euclid's Game calculates the greatest common divisor, in the next section we explore the fundamental idea of division in the integers.

3. The Division Algorithm

We know that the integer 6 divides 54 because $54 = 6 \cdot 9$. Consequently, 54 can be divided into 6 groups of 9. The number 57 is not divisible by 6, but it is quite useful to think of 57 as 6 groups of 9 with 3 left over, or, we say, with 3 as remainder. This is an example of what is known as *integer division with remainder*.

The idea of division in $\mathbb{Z}$ of any positive integer a by another positive integer b is just the same as in the example. Let's assume for the moment that $b < a$. Then we can divide a into a number, say q, of groups of b, where q is chosen so what's left over, the remainder, is a nonnegative integer less than b. Observe that if $b \mid a$, then there is 0 left over. We prove that it is always possible to execute division with remainder in this way.

The Division Algorithm. If a and b are positive integers, then there are unique nonnegative integers q and r such that
$$a = bq + r, \text{ with } 0 \leq r < b.$$

In the equation above, a is the *dividend*, b is the *divisor*, q is the *quotient* and r is the *remainder*.

Note that it is not the division itself, but the resriction $0 \leq r < b$ on the remainder that is the essence of the algorithm. We call on the Well Ordering Property to establish division with the prescribed remainder. Recall that the Well Ordering Property states that in every nonempty subset of the positive integers, there is a smallest positive integer.

Here are some examples of the division algorithm. If we divide 39 by 7 using the division algorithm, we have
$$39 = 7 \cdot 5 + 4,$$

where the dividend is 39, the divisor is 7, the quotient is 5 and the remainder is 4 and $4 < 7$.

If we use the division algorithm to divide 54 by 12 and to divide 153 by 17, we have
$$54 = 12 \cdot 4 + 6, \quad \text{and} \quad 153 = 17 \cdot 9 + 0.$$

The remainder is equal to 0 in the last example because $17 \mid 153$.

Verification of the Division Algorithm. First, we find integers q and r satisfying the requirements. Then we discuss what it means to say they are unique.

If $b > a$, a very uninteresting case where the divisor is larger than the dividend, we may write

$$a = b \cdot 0 + a,$$

where $q = 0$ and $r = a$. This case yields no useful information.

Suppose that $b \leq a$. If $b \mid a$, then a is a multiple of b, so we have

$$a = bq,$$

for some q, and $r = 0$.

Assume that $b < a$ and $b \nmid a$. We seek to split a into groups of b so that what is left over is less than b. In other words, we want to subtract a multiple bk of b from a in such a way that the remainder $a - bk$ is a nonnegative integer less than b. To see that this can be done, let us designate by S the set of all positive differences of the form a minus positive integer multiples of b :

$$S = \{a - bk \mid k \text{ in } \mathbb{N} \text{ and } a - bk > 0\}.$$

This is a set of positive integers because we have defined it to be so. Is it nonempty? Yes. Remember that $a > b$, so, if we take $k = 1$, we have that $a - b$, which is greater than 0, is in S. Thus, the requirements for the Well Ordering Property are in place. By that property, in S there is a least positive integer which we will call r. Since r is in S, r must be a positive difference of the form $r = a - bq$, for some q in $\mathbb{N}$.

We want to show that $r < b$. Suppose, instead, that $r \geq b$. First, note that $r \neq b$. For if $r = b$, then

$$a = bq + b.$$

So, by the distributive rule,

$$a = b(q + 1),$$

and $b \mid a$, contrary to our assumption that $b \nmid a$.

It remains to show that r cannot be greater than b. We show that if $r > b$, then there is an element in S that is less than r, the smallest element in S. The reason is that

$$r - b = a - bq - b.$$

Therefore, by the distributive rule,

$$r - b = a - b(q + 1).$$

Observe that the integer $r - b = a - b(q + 1)$ is a positive difference of the form a minus a positive multiple of b. Therefore, it is in S. But $b > 0$, so

$r - b$ is less than r, the least element in S. This contradiction means that our assumption that $r > b$ is false. Consequently, since we have ruled out the case $r = b$, it must be the case that r is less than b, which is what we desired to prove.

In the statement of the algorithm, the integers q and r are said to be unique. What does this mean? To say that a defined number, or quantity, is unique means that *there is only one number, or quantity, that satisfies the restrictions set out in its definition.* For example, there are forty eight integers greater than 1 and less than 50. However, there is only one integer greater than 1 and less than 3, namely 2. So, 2 is the unique integer greater than 1 and less than 3. For an example from geometry, consider that among all the lines from the vertex of a triangle to its base, there is a unique line from the vertex to the midpoint of the base.

✳

Seminar Activity.

  (i) Give an example of a concept that is not unique and then add restrictions to its definition so that it is unique.

 (ii) Find an example in print of the word *unique* attributed incorrectly.

✳

We return now to the statement of the division algorithm and focus on the uniqueness of r. Observe that when integer division is performed, many different remainders can be obtained. For example, if $a = 14$ and $b = 4$, we have, among many other expressions:

$$14 = 5 \cdot 2 + 4 \ \text{ and } \ 14 = 5 \cdot 3 - 1.$$

with remainders 4 and -1, respectively. Thus, the remainder obtained in integer division is *not* unique. What is true is that *if* the remainder r satisfies $0 \leq r < b$, as we have in the statement of the division algorithm, *then* r is unique. Moreover, when we require that $0 \leq r < b$, so that r is unique, then q is also unique. Thus, if the remainder r satisfies $0 \leq r < b$, there is only one way to carry out integer division and that way is described by the division algorithm.

The following observation will be applied frequently. It is the key to computing the greatest common divisor, for it allows us to swap one pair of numbers for another when calculating it. The observation states that the greatest common divisor of the dividend and the divisor is the greatest common divisor of the divisor and the remainder.

OBSERVATION 3.1. *Let a and b be positive integers with $a > b$. Suppose that $a = bq + r$, with $0 < r < b$. If $d = \gcd(a,b)$, then $d \mid r$ and $d = \gcd(b,r)$.*

✳

Seminar Exercise. Use the fact that d divides every linear combination of a and b to verify Observation 3.1.

✳

Comments on the Seminar Exercise. Here is one way to do the exercise. Since $r = a - bq$ is a linear combination of a and b, it follows, as we have observed many times, that $d \mid r$. To see that d is the largest integer dividing b and r, observe that if the positive integer c divides b and r, then $c \mid a = bq+r$, so $c \mid a$ and $c \mid b$. By definition of the greatest common divisor d of a and b, we have $c \mid d$, so, by Property (CCD) in Seminar 2, Section 6, $c \leq d$.

✳

Note. The division algorithm can be extended to hold for any integer dividend a. It is true that if a and b are integers with $b > 0$, then there are unique integers q and r such that

$$a = bq + r, \text{ with } 0 \leq r < b.$$

The verification of the algorithm that we presented assumes that $a > 0$, so a slightly different one must be given if we allow a to be negative.

4. Return to Euclid's Game

The division algorithm is exactly the tool needed to verify the essential fact about Euclid's Game, namely, that the game provides a straightforward (and fun) way to calculate the greatest common divisor of a and b.

OBSERVATION 4.1. *Let a and b be positive integers. The smallest integer obtained as a positive successive difference in Euclid's Game, played with a and b, is $\gcd(a,b)$.*

To verify the observation, we let E be the set of all of the numbers on the board at the finish of Euclid's Game. In other words, E is the set of positive integers consisting of a and b and of all successive differences obtained in the game. The set E is a nonempty set of positive integers. If you guessed that these words signal an application of the Well Ordering Property, you are correct. By that property, E has a smallest positive integer which we call d.

We claim that $d \mid a$ and $d \mid b$. To see this, we apply the division algorithm to a and d. The algorithm gives $a = dq + r$, where $0 \leq r < d$. To show that $d \mid a$, we need to show $r = 0$. If, on the contrary, $r > 0$, then $a - dq > 0$ and the positive successive differences $a - d$, $(a - d) - d = a - 2d$, $\dots$, $a - dq$ are numbers in E. But, $r = a - dq < d$. This contradicts the fact that d is the least element in E. Therefore, $r = 0$, and $d \mid a$. The fact that $d \mid b$ can be shown in a similar way. Since d is in E, we know that d is a linear combination of a and b. We have shown that $d \mid a$, $d \mid b$, and d is a linear combination of a and b. Consequently, by Observation 1.2, $d = \gcd(a,b)$.

Remark. In the discussion above, the division algorithm is applied to show that one number actually divides another, i.e., that the remainder is zero. This is a useful technique.

The next result is the most important and most frequently applied result about the greatest common divisor. We will use it often for practical and theoretical purposes. Observation 4.1 states that $\gcd(a,b)$ is the smallest integer in the set E consisting of the linear combinations obtained as positive successive differences in Euclid's Game. Thus the corollary below follows immediately.

COROLLARY 4.2. $\gcd(a,b)$ *is a linear combination of a and b.*

Recall that a pair of integers a and b are said to be *relatively prime* if $\gcd(a,b) = 1$. We have a notable special case of the previous corollary.

COROLLARY 4.3. *If a and b are relatively prime, then 1 can be written as a linear combination of a and b, that is, $1 = sa + tb$, for some integers s and t.*

When we play Euclid's Game to calculate $\gcd(a,b)$, it is helpful to know when all of the positive successive differences have been found. It is a fact that if $a > b$, the precise number of positive integers in E, the set consisting of a and b and all positive successive differences (these are the numbers written on the board) is

$$\frac{a}{\gcd(a,b)}.$$

Example. The numbers on the board when we play Euclid's Game with the positive integers 105 and 45, are

$$105, 45, 60, \text{ and, in some order, } 15, 30, 90, 75.$$

We see that $\gcd(105, 45) = 15$, the smallest number. Note that $a/(\gcd(a,b))$ $= 105/15 = 7$, and this is precisely the number of integers on the board. To write 15 as a linear combination of 105 and 45, we look at the successive differences found in the game. The gcd 15 is a second positive successive difference computed as follows: $(105 - 45) - 45 = 15$, so 15 can be written

$$15 = (1)105 + (-2)45,$$

a linear combination of 105 and 45 with coefficients $s = 1$ and $t = -2$. See Section 6 where several methods for finding the coefficient s and t are explained.

✳

Seminar/Classroom Activity. Play Euclid's Game with 231 and 126. Find $\gcd(231, 126)$. Write $\gcd(231, 126)$ as a linear combination of 231 and 126. Show that the number of integers on the board, i.e., the number of successive differences including 231 and 126, is

$$\frac{231}{\gcd(231, 126)}.$$

❋

Comments on the Seminar/Classroom Activity. When you play the game, you will find that $\gcd(231, 126) = 21$. The numbers on the board when the game is finished are (in some order):

$$231,\ 126,\ 210,\ 189,\ 168,\ 147,\ 105,\ 84,\ 63,\ 42,\ 21.$$

We see that $231 - 126 = 105$ and $126 - 105 = 21$. Thus, $21 = 2 \cdot 126 + (-1) \cdot 231$. There are 11 numbers on the board, and $11 = 231/21$.

❋

The fact that $\gcd(a, b)$ is a linear combination of a and b is *a very significant result in the theory of numbers in* $\mathbb{Z}$. In the next section, we introduce the Euclidean algorithm, *the most efficient method for computing* $\gcd(a, b)$.

5. The Euclidean Algorithm

Euclid's algorithm is a technique for finding $\gcd(a, b)$ for any positive integers a and b. We may assume that $a > b$, since $\gcd(a, b) = \gcd(b, a)$. The Euclidean algorithm is based on repeated applications of the division algorithm. Most probably, the process will have several steps, so we index the quotients and remainders that arise by the step number. We will apply Observation 3.1 frequently. That observation states that if $a = bq + r$, with $0 \leq r < b$ and if $d = \gcd(a, b)$, then $d = \gcd(b, r)$.

We will see how this result allows us to put off our calculation of the gcd until we have an equation where the gcd is obvious.

The Euclidean Algorithm. Let a and b be positive integers with $a > b$. We apply the division algorithm to divide a by b :

$$a = bq_1 + r_1, \text{ where } 0 \leq r_1 < b.$$

If $r_1 = 0$, then $a = bq_1$ is a multiple of b, so $d = \gcd(a, b) = b$ and the algorithm ends in one step.

If $r_1 > 0$, it follows from Observation 3.1 that $d = \gcd(b, r_1)$. Since $b > r_1$, we apply the division algorithm to divide b by r_1 :

$$b = r_1 q_2 + r_2, \text{ where } 0 \leq r_2 < r_1.$$

If $r_2 = 0$, then b is a multiple of r_1, so $d = \gcd(a, b) = \gcd(b, r_1) = r_1$, and the algorithm ends in two steps.

If $r_2 > 0$, then it follows from Observation 3.1 that $d = \gcd(a, b) = \gcd(b, r_1)$ $= \gcd(r_1, r_2)$. Since $r_1 > r_2$, we use the division algorithm to divide r_1 by r_2 :

$$r_1 = r_2 q_3 + r_3, \text{ where } 0 \leq r_3 < r_2.$$

If $r_3 = 0$, then r_1 is a multiple of r_2, so $\gcd(r_1, r_2) = r_1$, and $d = \gcd(a, b) = \gcd(b, r_1) = \gcd(r_1, r_2) = r_1$, and the algorithm ends in three steps. Note that we have

$$0 \leq r_3 < r_2 < r_1 < b.$$

As long as the remainder is nonzero, we may apply the division algorithm and obtain a smaller remainder. We claim that there is a smallest numbered step, say step $n + 1$, where the remainder must be 0. Otherwise, the set of positive integer remainders keeps decreasing, is nonempty, and has no least element, a contradiction to the Well Ordering Property of the positive integers.

Thus, we have at
step n,

$$r_{n-2} = r_{n-1}q_n + r_n, \text{ where } 0 < r_n < r_{n-1},$$

and at
step n + 1,

$$r_{n-1} = r_n q_{n+1} + 0.$$

Moreover, it follows from Observation 3.1 that

$$d = \gcd(a, b) = \gcd(b, r_1) = \gcd(r_1, r_2) = \cdots$$
$$= \gcd(r_{n-1}, r_n) = \gcd(r_n, 0) = r_n.$$

Consequently, we have shown that **$\gcd(a, b)$ is the last nonzero remainder.**

We illustrate how the algorithm works with the following example. We find $\gcd(343, 280)$ by applying the algorithm. In the first step, the dividend is 343 and the divisor is 280:

$$343 = 280 \cdot 1 + 63.$$

For the second step, 280, the divisor above, becomes the dividend, and 63, the remainder above, becomes the divisor:

$$280 = 63 \cdot 4 + 28.$$

For the third step, 63, the divisor above, becomes the dividend, and 28, the remainder above, becomes the divisor:

$$63 = 28 \cdot 2 + 7.$$

For the fourth step, 28, the divisor above, becomes the dividend, and 7, the remainder above becomes the divisor:

$$28 = 7 \cdot 4 + 0.$$

The last nonzero remainder is 7, so $\gcd(343, 280) = 7$.

✳

Seminar Activity. As we have observed, the divisor in step i of the Euclidean Algorithm becomes the dividend in step $i+1$, and the remainder in step i, if nonzero, becomes the divisor in step $i+1$. Illustrate this by drawing an arrow from divisor to dividend and from remainder to divisor in each step of the algorithm listed below. This exercise will help you remember how to do the algorithm.

$$a = bq_1 + r_1, \text{ where } 0 \leq r_1 < b.$$

$$b = r_1 q_2 + r_2, \text{ where } 0 \leq r_2 < r_1.$$

$$r_1 = r_2 q_3 + r_3, \text{ where } 0 \leq r_3 < r_2$$

$$r_2 = r_3 q_4 + r_4, \text{ where } 0 \leq r_4 < r_3$$

$$r_3 = r_4 q_5 + r_5, \text{ where } 0 \leq r_5 < r_4$$

$$r_4 = r_5 q_6 + r_6, \text{ where } 0 \leq r_6 < r_5$$

$$\vdots$$

$$r_{n-3} = r_{n-2} q_{n-1} + r_{n-1}, \text{ where } 0 \leq r_{n-1} < r_{n-2}$$

$$r_{n-2} = r_{n-1} q_n + r_n, \text{ where } 0 \leq r_n < r_{n-1}$$

$$r_{n-1} = r_n q_{n+1} + 0$$

✳

Example. We calculate $\gcd(78696, 19332)$.

$$78696 = 19332 \cdot 4 + 1368$$

$$19332 = 1368 \cdot 14 + 180$$

$$1368 = 180 \cdot 7 + 108$$

$$180 = 108 \cdot 1 + 72$$

$$108 = 72 \cdot 1 + 36$$

$$72 = 36 \cdot 2 + 0$$

Thus the last nonzero remainder is 36, so $\gcd(78696, 19332) = 36$.

✳

Seminar Exercise. Use the Euclidean Algorithm to calculate $\gcd(56, 24)$, $\gcd(91, 85)$, $\gcd(288, 156)$ and $\gcd(10000, 875)$.

✳

Comments on the Seminar Exercise. By applying the Euclidean Algorithm, we obtain $\gcd(56, 24) = 8$, $\gcd(91, 85) = 1$, $\gcd(288, 156) = 12$ and $\gcd(10000, 875) = 125$.

The steps of the Euclidean Algorithm to calculate $\gcd(91, 85)$ are:

$$91 = 85 \cdot 1 + 6$$
$$85 = 6 \cdot 14 + 1$$
$$6 = 1 \cdot 6 + 0$$

Thus $\gcd(91, 85) = 1$.

✳

6. GCD's as Linear Combinations

As we have mentioned several times, the fact that the positive integer $d = \gcd(a, b)$ is a linear combination of a and b is one of the most significant facts about the greatest common divisor. We assume throughout this section that $d \neq a$ and $d \neq b$.

We know that to say $d = \gcd(a, b)$ is a linear combination of a and b means that there are integers s and t, where s and t may be positive or negative, such that

$$d = sa + tb.$$

The values of s and t such that $\gcd(a, b) = sa + tb$ are not unique. It is always wise to first inspect d, a and b to check for obvious values for s and t.

The question we answer in this section is: How do we find values for s and t? We assume that $d = \gcd(a, b)$ has been calculated, by inspection or by the Euclidean Algorithm. We present three methods for computing s and t.

(i) **(CA) The Coefficient Algorithm.** We assume, as we may, that $a < b$, and we explain how to calculate integers s and t, such that

$$d = sa + tb.$$

Step 1. Divide the equation $d = sa + tb$ through by d. The equation that results is

$$1 = sa' + tb', \quad \text{where} \quad a' = \frac{a}{d} \quad \text{and} \quad b' = \frac{b}{d}$$

are integers. (See Observation 1.1.)

Step 2. Rewrite the equation above in the form:

$$sa' = 1 - tb'.$$

Step 3. Choose and test values of t from $-b'/2$ to $b'/2$, if b' is even, and from $-(b'-1)/2$ to $(b'-1)/2$, if b' is odd, beginning with 1 until you find a value of t such that a' divides $1 - tb'$. The codivisor s of a' for this value of t, and t itself are the coefficients we are seeking.

value of t	$1 - b't$	Does a' divide $1 - b't$?
1	$1 - b'$	Does a' divide $1 - b'$?
-1	$1 - (-1)b'$	If $a' \nmid 1 - b'$, does a' divide $1 + b'$?
2	$1 - 2b'$	If $a' \nmid 1 + b'$, does a' divide $1 - 2b'$?
-2	$1 + 2b'$	If $a' \nmid 1 - 2b'$, does a' divide $1 + 2b'$?
$\vdots$	$\vdots$	$\vdots$

Step 4. If s and t denote the solutions found in Part 3, we mutiply the equation $1 = sa' + tb'$ through by d to obtain the equation

$$d = s(a'd) + t(b'd) = sa + tb,$$

as required.

You will appreciate how neat the algorithm is when we do an example. We note, however, that if a and b are very large, checking all the values of t in Step 3 above could be a daunting task.

Example. Let $a = 22$ and $b = 34$. We see that $\gcd(22, 34) = 2$, so there are integers s and t such that

$$22s + 34t = 2.$$

We seek s and t. The first step is to divide this equation through by 2. The result is the equation

$$11s + 17t = 1.$$

We apply the coefficient algorithm to calculate s and t as follows. We rewrite the equation above:

$$11s = 1 - 17t,$$

where $a' = 22/2 = 11$ and $b' = 34/2 = 17$. Since b' is odd, we look for a value of t between -8 and 8 so that 11 divides $1 - t \cdot 17$.

value of t	$1 - 17t$	11 divides $1 - 17t$?
1	$1 - 17 = -16$	No, $11 \nmid -16$
-1	$1 + 17 = 18$	No, $11 \nmid 18$
2	$1 - 17 \cdot 2 = -33$	Yes, $-33 = -3 \cdot 11$.

Thus, $t = 2$, $s = -3$, and

$$11(-3) + 17(2) = 1.$$

By multiplying this equation through by 2, we have $22(-3) + 34(2) = 2$.

Example. Let $a = 60$ and $b = 84$. By inspection, or the Euclidean Algorithm, we find that $\gcd(60, 84) = 12$.
We seek integers s and t so that

$$12 = s \cdot 60 + t \cdot 84.$$

Step 1. Divide the equation above through by 12.

$$1 = s \cdot 5 + t \cdot 7.$$

Step 2. Rewrite the equation.

$$s \cdot 5 = 1 - t \cdot 7.$$

Step 3. $b' = 7$ is odd. Consequently, we test values of t, from -3 to 3, beginning with $t = 1$, until a value of t is found so that $1 - t \cdot 7$ is divisible by 5.

value of t	$1 - t \cdot 7$	5 divides $1 - t \cdot 7$?
1	$1 - 7 = -6$	No, $5 \nmid -6$.
-1	$1 + 7 = 8$	No, $5 \nmid 8$
2	$1 - 2 \cdot 7 = -13$	No, $5 \nmid -13$
-2	$1 + 2 \cdot 7 = 15$	Yes, $15 = 3 \cdot 5$.

When $t = -2$, we have $1 + 2 \cdot 7 = 15$, and $15 = 3 \cdot 5$ is divisible by 5 with codivisor 3. Thus, $t = -2$ and $s = 3$, and

$$1 = 3 \cdot 5 + (-2) \cdot 7 = 3 \cdot 5 - 2 \cdot 7.$$

After we multiply through by 12, we have

$$12 = 3 \cdot 60 - 2 \cdot 84.$$

When you compare this method with the two others to follow, we think you will agree that it is the simplest and most direct.

✳

Seminar Exercise. Use the coefficient algorithm to calculate integers s and t so that $d = sa + tb$, for $a = 22$ and $b = 60$. Begin by calculating $\gcd(22, 60)$.

✳

Comments on the Seminar Exercise. By inspection (prime divisors) or by the Euclidean Algorithm, $\gcd(22, 60) = 2$. Thus,

$$2 = s \cdot 22 + t \cdot 60.$$

Step 1. Divide the equation above through by 2.

$$1 = s \cdot 11 + t \cdot 30.$$

Step 2. Rewrite the equation.

$$s \cdot 11 = 1 - t \cdot 30.$$

Step 3. Test values of t from $t = -15$ to $t = 15$, beginning with with $t = 1$, until a value is found such that $1 - t \cdot 30$ is divisible by 11.

value of t	$1 - t \cdot 30$	11 divides $1 - t \cdot 30$?
1	$1 - 30 = -29$	No, $11 \nmid -29$
-1	$1 + 30 = 31$	No, $11 \nmid 31$
2	$1 - 60 = -59$	No, $11 \nmid -59$
-2	$1 + 60 = 61$	No, $11 \nmid 61$
3	$1 - 90 = -89$	No, $11 \nmid -89$
-3	$1 + 90 = 91$	No, $11 \nmid 91$
4	$1 - 120 = -119$	No, $11 \nmid -119$
-4	$1 + 120 = 121$	Yes, $11 \cdot 11 = 121$

Thus, $t = -4$ and $s = 11$, and

$$1 = 11 \cdot 11 - 4 \cdot 30.$$

Step 4. When we multiply through by 2, we have

$$2 = 11 \cdot 22 - 4 \cdot 60.$$

✳

(ii) **(REA) The Reverse Euclidean Algorithm.** We show how to read the Euclidean Algorithm backwards to find integers s and t so that

$$d = sa + tb,$$

where $d = \gcd(a, b)$. Note that here we write d as a sum of two terms in the usual way. However, since d is a positive integer less than or equal to a and to b, it follows that one of the coefficients s and t is positive and one is negative.

We do not write down an algorithm for the general case where a and b are arbitrary integers. The equations that arise in the process of finding the integers s and t so that $\gcd(a, b) = sa + tb$ are formidable, and are not particularly useful. What is important to convey to students is that particular examples are not difficult to work out, but the process must be carried out meticulously. We illustrate this with three examples.

Example 1. For our first example of reversing the Euclidean Algorithm, we take one of the examples from the last Seminar Exercise in Section 5, $\gcd(56, 24) = 8$, and demonstrate how to find s and t so that $8 = \gcd(56, 24) = s \cdot 56 + t \cdot 24$. For this example, the Euclidean Algorithm has only two steps:

$$56 = 24 \cdot 2 + 8$$
$$24 = 8 \cdot 3 + 0.$$

To find the coefficients s and t, we **begin at the next to the last step in the algorithm, the step where the remainder is the greatest common divisor.** Our example has only two steps, so the next to the last step is the first step. We solve the equation $56 = 24 \cdot 2 + 8$ for 8, to obtain

$$8 = 56 + (-2) \cdot 24 = 1 \cdot 56 + (-2) \cdot 24.$$

We have $8 = \gcd(56, 24)$ written as a linear combination of 56 and 24 with coefficients $s = 1$ and $t = -2$.

Example 2. For the next example, we write $\gcd(48, 18) = 6$ as a linear combination of 48 and 18. Here are the steps of the Euclidean algorithm:

$$48 = 18 \cdot 2 + 12$$
$$18 = 12 \cdot 1 + 6$$
$$12 = 6 \cdot 2 + 0.$$

To find the coefficients s and t, we **begin at the next to the last step in the algorithm, the step where the remainder is the greatest common divisor.** This example has three steps, so the next to the last step is the second step. We solve the equation $18 = 12 \cdot 1 + 6$ for 6, to obtain

$$6 = 18 - 12 \cdot 1.$$

We have 6 written as a linear combination of 18 and 12. This is not quite what we need, but we see that in the first step we find 12 written as a linear combination of 48 and 18. So we solve this equation for 12 with the result $12 = 48 - 18 \cdot 2$. We substitute $48 - 18 \cdot 2$ for 12 in the displayed equation above to obtain

$$6 = 18 - (48 - 18 \cdot 2).$$

We regroup and find that

$$6 = (-1) \cdot 48 + (1 + 2) \cdot 18 = (-1) \cdot 48 + 3 \cdot 18.$$

We have $6 = \gcd(48, 18)$ written as a linear combination of 48 and 18 with coefficients $s = -1$ and $t = 3$.

Example 3. As a final example, we consider $\gcd(343, 280)$ computed in Section 5, and demonstrate how to find s and t so that $7 = \gcd(343, 280) = s \cdot 343 + t \cdot 280$. Here are the four steps of the Euclidean algorithm to compute $\gcd(343, 280)$ which we refer to as (EA) in the calculations we do below.

$$343 = 280 \cdot 1 + 63$$
$$280 = 63 \cdot 4 + 28$$
$$63 = 28 \cdot 2 + 7$$
$$28 = 7 \cdot 4 + 0$$

We seek integers s and t so that

$$7 = s \cdot 343 + t \cdot 280.$$

To find them, we **begin at the next to the last step in (EA), the step where the remainder is the greatest common divisor.** In our example, it is the third equation in (EA), where $7 = \gcd(343, 280)$ is the remainder. So we solve this equation for 7, and obtain

$$7 = 63 + (-2) \cdot 28.$$

We proceed from the next to the last up to the first of the equations substituting the remainders each step of the way, applying the distributive rule and regrouping as we go. For the next step, we solve the second equation in (EA) for 28, substitute it in the equation above, apply the distributive rule, regroup and find that

$$28 = 280 + (-4)63.$$

Thus,

$$7 = 63 + (-2)28$$
$$7 = 63 + (-2)[280 + (-4)63]$$
$$7 = (1 + (-2)(-4))63 + (-2)280$$
$$7 = 9 \cdot 63 + (-2)280.$$

You see that we have 7 written as a linear combination of 63 and 280, so, as the final step, we solve the first equation in (EA) for 63, substitute in the last equation above, apply the distributive rule and regroup. We have

$$63 = 343 + (-1)280.$$

We substitute the expression on the right above for 63 in the equation

$$7 = 9(63) + (-2)280,$$

apply the distributive rule and regroup to obtain

$$7 = 9[343 + (-1)280] + (-2)280$$
$$7 = 9 \cdot 343 + (-11) \cdot 280.$$

Thus, we have $\gcd(343, 280) = 7$ written as a linear combination:

$$7 = 9 \cdot 343 + (-11) \cdot 280.$$

Always check, by multiplication and addition, that the linear combination you find is correct.

The process of substitution, applying the distributive rule and regrouping is valuable experience for students. The computations must be done carefully.

(iii) **(REG) Reverse Euclid's Game.** This method is analogous to the Reverse Euclidean Algorithm. Again, there is no succinct algorithm that is easily written down for the general case. Particular examples are not difficult to work out, but the process must be carried out carefully. We write out only one example since the process is so similar to (REA).

Example. We consider Euclid's Game in Section 2 where $a = 280$ and $b = 105$. We reprise the steps of that example where we found that $\gcd(a, b) = 35$.

Player A	Player B	Numbers on the Board
$280 - 105 = 175$		$280, 105, 175$
	$175 - 105 = 70$	$280, 105, 175, 70$
$105 - 70 = 35$		$280, 105, 175, 70, 35$
	$175 - 35 = 140$	$280, 140, 105, 175, 70, 35$
$280 - 35 = 245$		$280, 245, 175, 140, 105, 70, 35$
	$245 - 35 = 210$	$280, 245, 210, 175, 140, 105, 70, 35$

We look for the first play of the game where the gcd, 35, appears. In this example, it is the third play of the game. In the third play, Player A takes the difference

$$105 - 70 = 35,$$

where 70 is the difference $70 = 175 - 105$ that Player B finds in the second play. Thus, by the distributive rule,

$$35 = 105 - 70 = 105 - (175 - 105) = 2(105) - 175.$$

But, 175 is the difference $175 = 280 - 105$ Player A finds in the very first play. It follows, again by the distributive rule, that

$$35 = 2(105) - 175$$
$$= 2(105) - (280 - 105)$$
$$= 3(105) - 280.$$

Consequently, we can write $35 = s(280) + t(105)$, where $s = -1$ and $t = 3$.

✳

Seminar Exercise. In one of the exercises in Section 4, you played Euclid's Game with 231 and 126 and calculated $\gcd(231, 126)$. Use (REG) to write $\gcd(231, 126)$ as a linear combination of 231 and 126.

✳

Comments on the Seminar Activity. When you play the game, you will find that $\gcd(231, 126) = 21$. The numbers on the board when the game is finished are (in some order):

$$231, \ 126, \ 210, \ 189, \ 168, \ 147, \ 105, \ 84, \ 63, \ 42, \ 21.$$

When you reverse the steps you took to find $\gcd(231, 126)$, you will see that $s = 2$ and $t = -1$. With hindsight after the game, you might see, more directly, that $231 - 126 = 105$ and $126 - 105 = 21$. Thus, $21 = 2 \cdot 126 + (-1) \cdot 231$.

✳

Prime Numbers and Factorization Into Primes

1. Prime Numbers

Every positive integer n greater than 1 has at least two positive divisors, for the equation $n = n \cdot 1$ implies that 1 and n divide n. In this seminar, we take a careful look at those integers greater than one that have *exactly* two positive divisors. These numbers are, as we shall see, the building blocks of the set of integers. They are called prime numbers. A *prime number* (often, *prime*, for short) is a positive integer greater than 1 with precisely two positive divisors, 1 and itself.

Since this seminar focuses on the positive divisors of positive integers, from now on, unless specifically stated otherwise, we use the word "divisor" to mean "positive divisor." By Property (CDD), we have that every divisor of a positive integer n is less than or equal to n.

Examples of "small" primes are easy to find. The only positive integers less than or equal to 2 are 1 and 2, so 2 is a prime number. The integers less than or equal to 3 are 1, 2 and 3. Since $3 = 2 + 1$, is odd, $2 \nmid 3$, so 3 is a prime number. What about 4, or for that matter, any even number greater than 2? Since every even number greater than 2 is divisible by 1, 2 and itself, no even number greater than 2 can be a prime number. A positive integer n greater than 1 that is not prime is called a *composite integer*. In other words, a positive integer n is composite if $n = st$, where s and t are integers satisfying $2 \le s < n$ and $2 \le t < n$. For example, as we have just observed, every even integer greater than 2 is a composite integer.

We will develop more tools for recognizing prime numbers shortly, but there is much to be learned from the simple activity below.

✻

Seminar/Classroom Activity. Identify each integer n from 2 to 25 as prime or composite, and justify your answers.

✻

Comments on Seminar and Classroom Activity. The positive integers
2, 3, 5, 7, 11, 13, 17, 19 and 23 are prime numbers. If n is any one of the
positive integers above, it is staightforward to check that none of the integers
k with $1 < k < n$ divides n. The remaining numbers are composite, since all
but four of them are even and, for those four, we have: $9 = 3 \cdot 3$, $15 = 3 \cdot 5$,
$21 = 3 \cdot 7$ and $25 = 5 \cdot 5$.

✳

(UTRP) Useful Tool For Recognizing Primes. Recall that if the
positive integer $n = st$, we regard s and t as codivisors of n. We claim that
$s^2 \leq n$ or $t^2 \leq n$. For if $s^2 > n$ and $t^2 > n$, then, by (O3),

$$n^2 = (st)^2 = s^2 t^2 > n \cdot n = n^2$$

Thus,

$$n^2 > n^2,$$

an impossibility. Thus, the square of one of every pair of codivisors of n is
less than or equal to n. Consequently,
**To test whether a number n is prime or not, it is sufficient to look
for divisors c with $c^2 \leq n$.** Put another way, *to test whether a number n
is prime or not, it is sufficient to look for divisors c with $c \leq \sqrt{n}$.*
Examples.
(i) Let us see if 47 is a prime number.
Since $7^2 = 49 > 47$, it is sufficient to check whether any positive integers,
just the odd ones are enough, less than 7 divide 47. You will find that none
do, so 47 is prime.

(ii) We check whether 167 is a prime number.
We have $169 = (13)^2$ and $169 > 167$, so it is sufficient to check whether any
odd integers less than 13 divide 167. It is straightforward to see that none
of the integers 3, 5, 7, 9, or 11 divides 167. (Observe $3 \nmid 167$, implies that
$9 \nmid 167$. Why?)

✳

Seminar/Classroom Activity. Test whether the integers 89 and 117 are
prime or not.

✳

Comments on the Seminar/Classroom Activity. The integer 89 is
prime. Since $(10)^2 = 100 > 89$, it follows that only the integers less than
10 need to be tested as possible divisors of 89. It is straightforward to check
that none do.
For 117, we have $(11)^2 = 121 > 117$. Thus, we only need to test the positive
integers less than 11 to see if any are divisors of 117. We find that 3 is a
divisor of 117, so 117 is not prime.

✳

A prime number that divides an integer n is called a *prime divisor* of n. For example, the prime numbers 3 and 13 are prime divisors of 117.

To help us identify larger numbers as prime or composite, we examine closely the properties of divisors of an arbitrary composite integer n. What can we say about the divisors of n that are greater than 1 and less than n?

We have the following significant observation.

OBSERVATION 1.1. *Every positive integer greater than 1 has a prime divisor, i.e., a divisor that is a prime number.*

Let n be a positive integer greater than 1. To verify this observation, we will call upon the Well Ordering Property which, as you know, says that there is a smallest positive integer in every nonempty set of positive integers. Let

$$S = \{\text{divisors of } n \text{ that are greater than 1.}\}$$

Observe that n is in S, for n is greater than 1 and is a divisor of n. Thus, S is a nonempty set of positive integers. We want to show that there is a prime number in the set S.

By the Well Ordering Property, S has a smallest positive integer that we will call k. If k is prime, then k is the desired prime divisor of n. Otherwise, k is a composite integer and can be written

$$k = st, \text{ where } 1 < s < k \text{ and } 1 < t < k.$$

But $k = st$ is in S, so k is a divisor of n, and every divisor of k is a divisor of n. It follows that both s and t must be in the set S. However, both s and t are less than k, the least element of S. This contradiction means that k must be prime.

You should try diligently to understand the statement in Observation 1.1. It, as well as the statement in the following corollary, are fundamental results that will be applied frequently.

COROLLARY 1.2. *Let n be a composite integer. Then n has a prime divisor p with $p^2 \leq n$, i.e., with $p \leq \sqrt{n}$.*

The reason is this. We know, recall (UTRP), that a composite number n has a divisor s with $s^2 \leq n$. By the observation above, s has a prime divisor p. Since $p^2 \leq s^2 \leq n$, it follows that $p^2 \leq n$, i.e., $p \leq \sqrt{n}$.

It follows from Corollary 1.2 that to test whether a positive integer n is prime or not, we only have to check *prime* numbers p with $p^2 \leq n$ to see if any divide n. For example, in the last Seminar Exercise you found that 89 is prime by testing for divisors among the integers from 2 to 10. However, now, by Corollary 1.2, only the *prime* numbers 2, 3, 5, and 7 from 2 to 10 need to be tested.

✳

Seminar Question. How many primes are there? Why do you think so?

✳

Comments on the Seminar Question. Since the positive integers become larger and larger, and since every positive integer has a prime divisor, it seems likely that there are infinitely many primes. The next observation demonstrates that this is so.

✳

OBSERVATION 1.3. *There are infinitely many prime numbers.*

Suppose, instead, that there are only a finite number of primes. It follows that we can make a list of *all* of them:

$$p_1, p_2, p_3, \ldots, p_k.$$

Consider the positive integer obtained by multiplying all the prime numbers on our list together and adding 1 :

$$s = p_1 p_2 p_3 \cdots p_k + 1.$$

We know, by Observation 1.1, that s is divisible by a prime number. But, by our assumption, all the prime numbers are on our list. So which prime on our list is it that divides s? If $p_1 \mid s$, then, $p_1 \mid s$ and $p_1 \mid p_1 p_2 p_3 \cdots p_k$, implies that

$$p_1 \mid (s - p_1 p_2 p_3 \cdots p_k).$$

However, $s - p_1 p_2 p_3 \cdots p_k = 1$, so $p_1 \mid 1$, an impossibility. The same argument can be applied to $p_2, p_3, \ldots, p_k$ to show that none of the primes on our list divides s. Thus our assumption is false and there are infinitely many primes.
Note. This proof was included in Euclid's *Elements* over 2000 years ago.

✳

Seminar/Classroom Activity. It is important to note that the argument above does *not* show that the number $s = p_1 p_2 p_3 \cdots p_k + 1$ is a prime number. It shows only that s is not divisible by any of the primes p_1, p_2, p_3, $\ldots$, p_k. Sometimes the number s is prime, sometimes not. Consider the primes 2, 3, 5, 7. Compute the numbers $2 \cdot 3 + 1$, $2 \cdot 3 \cdot 5 + 1$, and $2 \cdot 3 \cdot 5 \cdot 7 + 1$. Are these numbers prime?

✳

Comments on the Seminar and Classroom Activity. Yes, the numbers $7, 31$ and 211 are primes. The smallest number of the form $2 \cdot 3 \cdot 5 \cdots p_{t-1} \cdot p_t + 1$ that is not prime is $2 \cdot 3 \cdot 5 \cdot 7 \cdot 11 \cdot 13 + 1 = 30031 = 59 \cdot 509$. This example illustrates that testing just a few cases can lead to false conclusions.

✳

We end this section with a neat method for finding all the primes less than or equal to a specific integer n.

✳

Seminar/Classroom Activity. The *Sieve of Eratosthenes* is a method to find all the primes less than or equal to a fixed number. In this activity, we will find all of the primes less than or equal to 100.

As we have seen, by Corollary 1.2, every composite integer less than 100 has a prime factor less than $\sqrt{100} = 10$. These are the primes 2, 3, 5 and 7. Testing to see which of the numbers from 2 to 100 are divisible by the primes 2, 3, 5 and 7 is not difficult.

(i) Make a table of the numbers from 2 to 100.

	2	3	4	5	6	7	8	9	10
11	12	13	14	15	16	17	18	19	20
21	22	23	24	25	26	27	28	29	30
31	32	33	34	35	36	37	38	39	40
41	42	43	44	45	46	47	48	49	50
51	52	53	54	55	56	57	58	59	60
61	62	63	64	65	66	67	68	69	70
71	72	73	74	75	76	77	78	79	80
81	82	83	84	85	86	87	88	89	90
91	92	93	94	95	96	97	98	99	100

(ii) Cross out the integers in the table that are greater than 2 and divisible by 2.

	2	3	4̶	5	6̶	7	8̶	9	1̶0̶
11	1̶2̶	13	1̶4̶	15	1̶6̶	17	1̶8̶	19	2̶0̶
21	2̶2̶	23	2̶4̶	25	2̶6̶	2̶7̶	2̶8̶	29	3̶0̶
31	3̶2̶	33	3̶4̶	35	3̶6̶	37	3̶8̶	39	4̶0̶
41	4̶2̶	43	4̶4̶	45	4̶6̶	47	4̶8̶	49	5̶0̶
51	5̶2̶	53	5̶4̶	55	5̶6̶	57	5̶8̶	59	6̶0̶
61	6̶2̶	63	6̶4̶	65	6̶6̶	67	6̶8̶	69	7̶0̶
71	7̶2̶	73	7̶4̶	75	7̶6̶	77	7̶8̶	79	8̶0̶
81	8̶2̶	83	8̶4̶	85	8̶6̶	87	8̶8̶	89	9̶0̶
91	9̶2̶	93	9̶4̶	95	9̶6̶	97	9̶8̶	99	1̶0̶0̶

(iii) Continue sieving by crossing out the numbers greater than 3 that are divisible by 3, then the numbers greater than 5 that are divisible by 5 and, finally, finish sieving by crossing out the numbers greater than 7 that are divisible by 7. When you are done, the numbers remaining are all the primes less than or equal to 100.

✳

Comments on the Seminar/Classroom Activity. After you have crossed out all of the numbers not equal to 2, 3, 5 or 7 but divisible by 2, 3, 5 or 7, your table of primes less than or equal to 100 looks like this:

	2	3	4	5	6	7	8	9	10
11	12	13	14	15	16	17	18	19	20
21	22	23	24	25	26	27	28	29	30
31	32	33	34	35	36	37	38	39	40
41	42	43	44	45	46	47	48	49	50
51	52	53	54	55	56	57	58	59	60
61	62	63	64	65	66	67	68	69	70
71	72	73	74	75	76	77	78	79	80
81	82	83	84	85	86	87	88	89	90
91	92	93	94	95	96	97	98	99	100

Thus, the prime numbers less than or equal to 100 are

$$2, 3, 5, 7, 11, 13, 17, 19, 23, 29, 31, 37, 41, 43, 47, 53,$$
$$59, 61, 67, 71, 73, 79, 83, 89, 97.$$

Notice how the *theory* of numbers in $\mathbb{Z}$, in particular Corollary 1.2, made light work of compiling this list of primes. It is worth keeping for future reference.

*

Seminar/Classroom Activity.

(i) Make a Sieve of Eratosthenes to find all primes less than or equal to 500. (Save it for future reference.)

(ii) The positive integers 2 and 3 form a pair of consecutive prime numbers. Are there any other pairs of consecutive primes? Justify your answer.

(iii) The numbers 3 and 5, and 11 and 13 form two pairs of primes with just one number between them. For this reason, they are called *twin primes*. Are there any other pairs of twin primes? Find all pairs of twin primes that are less than 500. (Use your sieve.)

(iv) Observe that the three consecutive odd numbers 3, 5 and 7 are primes. They form a *prime triple*.

    (a) Show that every integer greater than 2 can be written in one of the following ways: $3q$, or $3q + 1$, or $3q + 2$, for some positive integer q.

    (b) Show that there are no prime triples consisting of three consecutive odd prime integers other than the prime triple 3, 5, 7.

*

Comments on the Seminar/Classroom Activity. For (i), note that $\sqrt{500} < 23$. Thus, we take, in turn each of the primes $p = 2, 3, 5, 7, 11, 13,$ 17 and 19, and cross out the numbers that are greater than p and divisible by p.

The answer to (ii) is "No," since 2 is the only even prime number.

For (iii), use the sieve you constructed in (i). The question of whether there are infinitely many pairs of twin primes is an open question at this time. The Twin Prime *Conjecture* states that there are an infinite number of pairs of twin primes.

For (a) in (iv), first observe that

$$3 = 3 \cdot 1, \quad 4 = 3 \cdot 1 + 1, \quad 5 = 3 \cdot 1 + 2, \quad 6 = 3 \cdot 2, \quad 7 = 3 \cdot 2 + 1.$$

To show that the statement is true in general, use the division algorithm. Let n be any integer greater than 2. Divide n by 3 :

$$n = 3q + r, \quad \text{where } q > 0 \text{ and } 0 \leq r < 3.$$

We take each of the three possibilities for r in turn. If $r = 0$, then $n = 3q$. If $r = 1$, then $n = 3q + 1$. Finally, if $r = 2$, then $n = 3q + 2$.

For (b) in (iv), suppose, on the contrary, that there is a triple p, $p + 2$, $p + 4$ of consecutive odd prime numbers with $p > 3$. By part (a), $p = 3q$, or $p = 3q + 1$, or $p = 3q + 2$, for some positive integer q. But p is a prime greater than 3, so $p \neq 3q$. If $p = 3q + 1$, then $p + 2 = 3q + 3 = 3(q + 1)$, so $p + 2$ is not prime. If $p = 3q + 2$, then $p + 4 = 3q + 6 = 3(t + 2)$ is not prime. Thus, in each case, we have shown that p, $p + 2$, $p + 4$ is not a prime triple.

✳

Here is a surprising fact about prime numbers that can be checked with a little work. **Interesting Fact About Prime Numbers.** If p is a prime number greater than 3, then p can be written in one of the following two ways:

$$p = 6m + 1 \text{ for some positive integer } m,$$

or

$$p = 6n - 1 \text{ for some positive integer } n.$$

For example, $5 = 6 \cdot 1 - 1$, $17 = 6 \cdot 3 - 1$, and $19 = 6 \cdot 3 + 1$. Observe that 17 and 19 form a pair of twin primes.
Warning. It is not true that, for every n, numbers of the form $6n + 1$ or $6n - 1$, must be prime numbers.
For example, $65 = 6 \cdot 11 - 1$ is divisible by 5, and is not a prime number.

To verify the Interesting Fact, we must show that every prime p greater than 3 can be written in the form $p = 6m + 1$, for some positive integer m, or $p = 6n - 1$, for some positive integer n. We noted above that $5 = 6 \cdot 1 - 1$. Consequently, we consider prime numbers p greater than 5. By Observation 1.3, there are infinitely many prime numbers, so there are, certainly, prime numbers greater than 5. Now, all positive integers $m > 5$ can be written in one of the following ways:

$$6n, \ 6n + 1, \ 6n + 2, \ 6n + 3, \ 6n + 4, \ \text{ or } 6n + 5$$

Why? This follows from applying the division algorithm to divide m by 6. We examine each of these types. Numbers of the form $6n$ are not prime because $6 \mid 6n$. Numbers of the form $6n + 2 = 2(3n + 1)$ are even and divisible by 2. Numbers of the form $6n + 3 = 3(2n + 1)$ are divisible by 3 and numbers of the form $6n + 4 = 2(3n + 2)$ are even and divisible by 2. Thus, if p is a prime number greater than 5, then $p = 6n + 1$ or $p = 6n + 5$. We may write

$$\begin{aligned} 6n + 5 &= 6n + 6 - 6 + 5 \\ &= 6(n + 1) - 1, \end{aligned}$$

so that a number that is an integer multiple of 6 plus 5 can always be written as a different integer multiple of 6 minus 1. Consequently, we have shown that if p is a prime number greater than 3, then p can be written in one of the following two ways:

$$p = 6m + 1 \text{ for some positive integer } m,$$

or

$$p = 6n - 1 \text{ for some positive integer } n.$$

⁂

Seminar/Classroom Activity. The purpose of this activity is to further your acquaintance with prime numbers. Your sieves will be very useful.

(i) Write each of the primes greater than 5 and less than 100 in the form $6n + 1$ or $6n - 1$.
(ii) Find the smallest integer of the form $6n+1$ or $6n-1$ that is not prime.
(iii) Find the smallest integer n such that neither of the integers $6n + 1$ or $6n - 1$ is prime.

Parts (ii) and (iii) suggest using the Interesting Fact to facilitate discovery of prime numbers. For numbers not of the form $6n - 1$ or $6n + 1$, for some n, can be rejected, and numbers of the form $6n - 1$ and $6n + 1$, for some n, can be tested.

⁂

2. Factorization Into Primes

In this section, we will study the essential idea that underlies a large part of the number theory of the integers. It is known as the Fundamental Theorem of Arithmetic. It states that every positive integer greater than 1 can be factored into a product of prime numbers, and that there is, basically, only one way to do this. For relatively small integers, such as 12, 55 and 138, factoring into primes is easy to check: $12 = 2 \cdot 2 \cdot 3$, $55 = 5 \cdot 11$, and $138 = 2 \cdot 3 \cdot 23$. We wish to show that *every* positive integer greater than 1 can be factored in this way. To accomplish this, it is necessary to investigate what happens when a prime divides a product of two or more numbers.

Think, for a moment, about numbers dividing products.

⁂

Seminar Question. Suppose a positive integer c divides a product ab of two positive integers a and b. Do you think that c must divide one of the factors a or b? Look at examples.

✳

Comments on the Seminar Question. The number 6 divides $8 \cdot 9 = 72$, but 6 does not divide either 8 or 9. Moreover 6 divides $2 \cdot 3$, but 6 does not divide either 2 or 3. In fact, every composite number c yields such an example. For we can write $c = ab$, where $1 < a < c$, $1 < b < c$. It follows that $c \mid ab$, but, by Property (CDD) in Seminar 2, Section 6, $c \nmid a$, and $c \nmid b$.

✳

A prime number p does not have divisors other than 1 and p. Perhaps p behaves differently when it divides a product. The next observation shows that it does. To verify it we use Seminar 3, Corollary 4.2 that states the important fact that the greatest common divisor of two numbers a and b can be written as a linear combination of a and b.

OBSERVATION 2.1 (**Euclid's Lemma**). *If a prime number p divides a product ab of two integers a and b, then it divides a or b (or both a and b).*

To verify the observation, we begin with our assumption that $p \mid ab$. If $p \mid a$, then we have the result we want and we are done. So, we assume that $p \nmid a$. We must show that $p \mid b$. Consider the greatest common divisor $\gcd(a, p)$. Since p has exactly two divisors 1 and p and our assumption is that $p \nmid a$, it follows that the only divisor common to a and p is 1. Thus, $\gcd(a, p) = 1$, and, by Seminar 3, Corollary 4.2, there are integers s and t so that

$$1 = sa + tp.$$

We multiply both sides of this equation by b :

$$b = sab + tpb.$$

We know $p \mid ab$ and it is clear that $p \mid tpb$, so, by the additive property of divisibility and the equation above, we have $p \mid (sab + tpb)$. But, this sum is equal to b. So, $p \mid b$, the desired conclusion.

There is another very useful result that extends the observation above to the case where we have an arbitrary positive integer, we will call it c, dividing a product ab, where c and a are relatively prime, i.e., $\gcd(c, a) = 1$.

OBSERVATION 2.2. *Let a, b and c be positive integers such that $\gcd(c, a) = 1$. If $c \mid ab$, then $c \mid b$.*

An argument analogous to that given for Observation 2.1 establishes this observation.

✳

Seminar Questions.

(i) Suppose p is a prime number that divides a product abc of three integers a, b and c. Do you think that p must divide a or b or c? Why do you think so?

(ii) Suppose p is a prime number that divides a product $a_1 a_2 a_3 a_4$ of four integers, or a product $a_1 a_2 a_3 \cdots a_k$ of k integers $a_1, \ldots, a_k$. Do you think that p must divide one of the factors a_i? Why do you think so?

✳

Comments on the Seminar Questions. The answer to the first question in part (i) and in part (ii) is "yes." Here is one way to think about them. A product $a_1 a_2 a_3 \cdots a_k$ of k integers can be regarded as a product of two integers, where the first integer is a_1 and the second integer is $a_2 a_3 \cdots a_k$:

$$a_1 a_2 a_3 \cdots a_k = a_1 \cdot (a_2 a_3 \cdots a_k).$$

Consequently, by Observation 2.1, p divides a_1 or p divides $a_2 a_3 \cdots a_k$. If p divides a_1, we are done. If not, then p divides $a_2 \cdot a_3 \cdots a_k = a_2 \cdot (a_3 \cdots a_k)$, so, again by Observation 2.1, p divides a_2 or p divides $a_3 \cdots a_k$, etc. You see how the argument goes. We arrive finally at the following corollary.

COROLLARY 2.3. *If a prime number p divides a product $a_1 a_2 a_3 \cdots a_k$ of k integers $a_1, \ldots, a_k$, then p divides (at least) one of the factors a_i.*

The following seminar exercise considers a special case of Corollary 2.3 where a prime p divides a product of primes.

✳

Seminar Exercise. If p and $q_1, q_2, q_3, \ldots, q_k$ are primes and if

$$p \mid (q_1 q_2 q_3 \cdots q_k),$$

we know, by Corollary 2.3, that p divides one of the primes q. It may be any one of the $q_1, q_2, q_3, \ldots, q_k$, but, by the commutative law for multiplication, we may assume that p divides q_1. Show that $p = q_1$.

✳

Comments on the Seminar Exercise. Our assumption is that p is a divisor of q_1. However, the prime q_1 has only two divisors, namely, 1 and q_1. Consequently, since $p \neq 1$, it follows that $p = q_1$.

✳

Remarks. We have the following conventions when we write products of primes. A single prime p is regarded as a product with one prime factor, namely p. A product of primes written as $p_1 p_2 p_3 \cdots p_t$, is not meant to suggest that the primes p_i are distinct. They need not be distinct. The factorization $4 = 2 \cdot 2$ is a simple example, and $50 = 2 \cdot 5 \cdot 5$ and $6615 = 3 \cdot 3 \cdot 3 \cdot 5 \cdot 7 \cdot 7$ are more robust examples of factorizations with repeated primes.

(FTA) The Fundamental Theorem of Arithmetic. Every positive integer $n > 1$ can be factored into a product of primes in exactly one way, up to order of the factors.

The phrase "in exactly one way, up to order of the factors" means

(i) there is precisely one set of primes that appear in the factorization of n;

(ii) if a prime p occurs k times in one factorization of n, then it must occur k times in every factorization of n;

(iii) the prime factors may be written down in different orderings.

For example, we can shuffle the order of the primes in the facorization of $12 = 2 \cdot 2 \cdot 3$ to obtain $12 = 2 \cdot 3 \cdot 2$ and $12 = 3 \cdot 2 \cdot 2$, but every factorization has two 2's and one 3. Usually, we will write factorizations from left to right, smallest prime factors to largest.

We focus first on showing that n has a factorization into a product of prime numbers. Although the argument takes a few paragraphs to write down, the basic idea is to use Observation 1.1 repeatedly to "factor out" a prime number as many times as necessary until 1 is obtained as a codivisor.

Claim. Every positive integer $n > 1$ can be factored into a product of primes.

Verification of Claim. By Observation 1.1, n has a prime divisor so we may write:

$$n = p_1 n_1,$$

where p_1 is a prime and $n_1 \geq 1$. If $n_1 = 1$, then $n = p_1$, and the claim is verified. If $n_1 > 1$, then by Observation 1.1, n_1 has a prime divisor and we may write

$$n_1 = p_2 n_2,$$

where p_2 is a prime and $n_2 \geq 1$. If $n_2 = 1$, then $n = p_1 p_2$, and the verification is complete. Otherwise, by Observation 1.1, we may write

$$n_2 = p_3 n_3,$$

where p_3 is a prime and $n_3 \geq 1$. If $n_3 = 1$, then $n = p_1 p_2 p_3$, and the claim is verified. If not, the process may be continued, but not forever. For we have $n > n_1 > n_2 > n_3$, and every step of the process leads either to 1 or to a smaller positive integer. Here is where the Well Ordering Property can help us finish the proof. By that property, in every nonempty set of positive integers there is a smallest integer. Consequently, the nonempty set of all the codivisors

$$\{n_1, n_2, n_3, \ldots, \}$$

that arise in the factoring argument above has a smallest element which we call n_h. We claim that $n_h = 1$. Otherwise, by Observation 1.1, it has a prime divisor whose codivisor would be smaller than n_h, a contradiction. So $n_h = 1$, and

$$n = p_1 p_2 p_3 \cdots p_h.$$

For example, $252 = 2 \cdot 2 \cdot 3 \cdot 3 \cdot 7$, and $4679675 = 5 \cdot 5 \cdot 7 \cdot 11 \cdot 11 \cdot 13 \cdot 17$.

We verify the "uniqueness part" of (FTA) this way. We assume that an integer $n > 1$ has two factorizations

$$n = p_1 p_2 \cdots p_h = q_1 q_2 \cdots q_k$$

into primes $p_1, p_2, \ldots, p_h$ and primes $q_1, q_2 \ldots, q_k$, and we show that the number of primes in the factorization is the same (i.e., $h = k$) and each of the primes p is equal to some prime q. The key to the verification of this part of the Fundamental Theorem is the result from the Seminar Exercise preceeding the statement of (FTA). For, the equation

$$n = p_1 p_2 \cdots p_h = q_1 q_2 \cdots q_k$$

shows that p_1 divides the product $q_1 q_2 \cdots q_k$, thus it follows from Corollary 2.3 that p_1 divides one of the $q_1, q_2 \ldots, q_k$, which we may assume, by the commutativity property of mutiplication, is q_1. By the previous exercise,

$$p_1 \mid q_1 \text{ implies } p_1 = q_1.$$

Consequently,

$$p_1 p_2 \cdots p_h = p_1 q_2 \cdots q_k,$$

so, by multiplicative cancellation, it follows that

$$p_2 \cdots p_h = q_2 \cdots q_k.$$

We may apply the same argument repeatedly to show that every prime p on the left of the equation $p_2 \cdots p_h = q_2 \cdots q_k$ is equal to some prime q on the right. This means that the number h of p's is less than or equal to the number k of q's. But if $h < k$, then, after we cancelled all the p's on the left with the corresponding q's on the right, we would have an equation of the form 1 equals a product of q's, an impossibility. Thus, $h = k$ and our verification of (FTA) is complete.

Note. If the same prime appears more than once in a prime factorization, for example, as in

$$2100 = 2 \cdot 2 \cdot 3 \cdot 5 \cdot 5 \cdot 7,$$

then we will use exponential notation:

$$2100 = 2^2 \cdot 3 \cdot 5^2 \cdot 7.$$

As you see, the exponent of each prime is equal to the number of repetitions of that prime. The factorization of an integer $n > 1$ written as a product of *powers of distinct prime numbers* is called the *prime power factorization* of n. We will consistently use this form of (FTA). If p_1, p_2, $\ldots$, p_h are the distinct primes in the prime factorization of an integer $n > 1$, then we write

$$n = p_1^{a_1} p_2^{a_2} \cdots p_h^{a_h},$$

where the exponents a_1, a_2, $\ldots$, a_h are positive integers. Later, in certain special circumstances, it will be helpful to indicate that a particular prime, say q, is missing from the factorization. To do this, we write q^0. (Recall that $q^0 = 1$.) For example, in Seminar 5, we will write that $2100 = 2 \cdot 2 \cdot 3 \cdot 5 \cdot 5 \cdot 7$ is divisible by

$$2^0 \cdot 3 \cdot 5^2 \cdot 7^0 = 3 \cdot 5 \cdot 5 = 75.$$

How do we find the factorization of an integer $n > 1$? If n is large, it can be a difficult problem. This is why many secret codes and computer encryptions are based on prime factorizations. If the factorization cannot be found directly by inspection, try dividing n by prime numbers p, with $p^2 < n$, until you find a prime p that divides n. Then, apply the same process to the codivisor n_1 of p, and repeat.

For example, 515 is clearly divisible by 5. We have $515 = 5 \cdot 103$. Next we look for prime divisors of 103 among the primes p with p^2 less than or equal to 103. There are only four primes, 2, 3, 5 and 7, with squares less than or equal to 103, and none of them divides 103, so 103 is a prime (as you found earlier by the sieve method.) Consequently, the prime power factorization of 515 is

$$515 = 5 \cdot 103.$$

Consider $n = 252$. Obviously, 252 is even, so $252 = 2 \cdot 126$. The codivisor 126 is also even, so $126 = 2 \cdot 63$. Consequently, $252 = 2^2 \cdot 63$. The codivisor 63 is divisible by 3^2, and $63 = 3^2 \cdot 7$. Thus, the prime power factorization of 252 is

$$252 = 2^2 \cdot 3^2 \cdot 7.$$

✳

Seminar Exercise. Find the prime power factorization for each of the following integers: 180, 759, 1111, 841 and 4200.

✳

Comments on the Seminar Exercise. $180 = 2^2 \cdot 3^2 \cdot 5$, $759 = 3 \cdot 11 \cdot 23$, $1111 = 11 \cdot 101$, $841 = 29^2$, and $4200 = 2^3 \cdot 3 \cdot 5^2 \cdot 7$.

Seminar 5

Applications of Prime Power Factorization

1. Finding All Positive Divisors of n

This seminar focuses on interesting and useful applications of the prime power factorization of an integer $n > 1$. (In practical situations, it may be difficult to write down the prime factorization, but we know it exists.)

Recall that if h is an integer that divides n, then, there is an integer k, such that $n = hk$. The integer k also divides n, and h and k are called codivisors of n. If $h = k$, then $n = h^2$. As we succinctly summarized in Seminar 2, **divisors come in pairs, except for perfect squares.**

Since a prime dividing a product must divide one of the factors, it follows that a prime p divides an integer $n > 1$ precisely when it is equal to one of the prime factors of n. (See Seminar 4, Observation 2.1, Corollary 2.3, as well as the Seminar Exercise that follows it.)

With these facts in mind, we examine the prime power factorizations of divisors of n.

For example, the prime power factorization of 44 is

$$44 = 2^2 \cdot 11,$$

which has distinct primes 2, and 11. Thus, by Seminar 4, Observation 2.1, Corollary 2.3 and the following Seminar Exercise, there are only two primes which divide 44, namely 2, and 11. For the same reason, it follows that 2^2 is the highest power of 2 that divides 44. For if 2^3 divides 44, then, $2^3 \cdot h = 2^2 \cdot 11$, for some integer h. Then, by multiplicative cancellation, it follows that 2 divides 11, an impossibility. Consequently, the prime factors of a divisor of 44 are 2 and/or 11. Moreover, the positive divisors of 44 are all of the form

$$2^a \cdot 11^b,$$

where $0 \le a \le 2$, and $0 \le b \le 1$. For example, some of the divisors of 44 are

$$2^0 \cdot 11^0 = 1, \quad 2 \cdot 11^0 = 2, \quad \text{and } 2^2 \cdot 11 = 44.$$

Next, let us count the number of positive divisors of 44. To count, we will use over and over again the fact that the number of consecutive integers from

0 to any positive integer k is $k + 1$. For example, the number of consecutive integers from 0 to 3 is four. The consecutive integers are 0, 1, 2, and 3.

Counting the positive divisors of $44 = 2^2 \cdot 11$ means counting the number of positive integers of the form

$$2^a \cdot 11^b,$$

where $0 \le a \le 2$ and $0 \le b \le 1$. There are $2 + 1 = 3$ choices for a; they are 0, 1, and 2. There are $1 + 1 = 2$ choices for b; they are 0 and 1. To find the resulting number of all choices for the product $2^a \cdot 11^b$, we multiply the numbers 3 and 2 together to obtain 6. Thus, in all, there are

$$3 \cdot 2 = 6$$

positive divisors of 44. Note that the number of positive divisors of 44 is the product of $3 = 2 + 1$, and $2 = 1 + 1$.

To tabulate the divisors of 44, we let the exponent a run from 0 to 2 along the rows, and the exponent b run from 0 to 1 along the columns.

$2^0 11^0 = 1$	$2^1 11^0 = 2$	$2^2 11^0 = 4$
$2^0 11^1 = 11$	$2^1 11^1 = 22$	$2^2 11^1 = 44$

The reasoning supplied for the example 44 furnishes all divisors of any positive integer $n > 1$. We will summarize the general case at the close of this section.

Consider another example, the integer

$$200 = 2^3 \cdot 5^2.$$

The distinct primes appearing in its factorization are 2 and 5. The prime 2 has exponent 3 and the prime 5 has exponent 2.

✳

Seminar Question. How many positive divisors of 200 are there?

✳

Comments on the Seminar Question. We count the number of products $2^a \cdot 5^b$ where a can be any one of 0, 1, 2 or 3 and b can be any one of 0, 1 or 2. We have 4 choices of powers of 2 and 3 choices of powers of 5. To find all such products, therefore, we multiply 4 by 3 and find that the number of positive divisors of 200 is 12.

We display all of the positive divisors of 200 in a table. We know the positive divisors of 200 have the form $2^a \cdot 5^b$, where $0 \le a \le 3$, and $0 \le b \le 2$. In the table below, we let the exponent a run from 0 to 3 along the rows, and the exponent b run from 0 to 2 along the columns.

$2^0 5^0 = 1$	$2^1 5^0 = 2$	$2^2 5^0 = 4$	$2^3 5^0 = 8$
$2^0 5^1 = 5$	$2^1 5^1 = 10$	$2^2 5^1 = 20$	$2^3 5^1 = 40$
$2^0 5^2 = 25$	$2^1 5^2 = 50$	$2^2 5^2 = 100$	$2^3 5^2 = 200$

✳

Seminar/Classroom Activity.

(i) As you know
$$36 = 2^2 \cdot 3^2$$
is the prime factorization of 36. Write down all of the positive divisors of 36 in a table.

(ii) Do the same for the number
$$540 = 2^2 \cdot 3^3 \cdot 5.$$

(iii) Find the prime factorization of 954569. Compare it with the prime factorization of 540. Count the number of positive divisors of 954569. How does this number compare with the number of positive divisors of 540? What conclusions can you draw from these comparisons?

✳

Comments on the Seminar/Classroom Activity.

(i) We display the divisors of 36 in a table, as we did above for the integer 200.

$2^0 3^0 = 1$	$2^1 3^0 = 2$	$2^2 3^0 = 4$
$2^0 3^1 = 3$	$2^1 3^1 = 6$	$2^2 3^1 = 12$
$2^0 3^2 = 9$	$2^1 3^2 = 18$	$2^2 3^2 = 36$

(ii) This is a slightly more challenging problem because there are three distinct primes 2, 3 and 5.
$$540 = 2^2 \cdot 3^3 \cdot 5.$$
Let us count the number of divisors first. The divisors are all of the form
$$2^a \cdot 3^b \cdot 5^c.$$
There are 3 choices for a, namely 0, 1 and 2. There are 4 choices for b, namely 0, 1, 2 and 3. There are 2 choices for c, namely 0 and 1. Consequently, there are $3 \cdot 4 \cdot 2 = (2+1)(3+1)(1+1) = 24$ positive divisors of 540. We construct the table so that all of the divisors in the first four rows have the factor 5^0 and all of the divisors in the next four rows have the factor 5^1.

$2^0 3^0 5^0 = 1$	$2^1 3^0 5^0 = 2$	$2^2 3^0 5^0 = 4$
$2^0 3^1 5^0 = 3$	$2^1 3^1 5^0 = 6$	$2^2 3^1 5^0 = 12$
$2^0 3^2 5^0 = 9$	$2^1 3^2 5^0 = 18$	$2^2 3^2 5^0 = 36$
$2^0 3^3 5^0 = 27$	$2^1 3^3 5^0 = 54$	$2^2 3^3 5^0 = 108$
$2^0 3^0 5^1 = 5$	$2^1 3^0 5^1 = 10$	$2^2 3^0 5^1 = 20$
$2^0 3^1 5^1 = 15$	$2^1 3^1 5^1 = 30$	$2^2 3^1 5^1 = 60$
$2^0 3^2 5^1 = 45$	$2^1 3^2 5^1 = 90$	$2^2 3^2 5^1 = 180$
$2^0 3^3 5^1 = 135$	$2^1 3^3 5^1 = 270$	$2^2 3^3 5^1 = 540$

(iii) $954569 = 7^3 \cdot 11^2 \cdot 23$, and $540 = 2^2 \cdot 3^3 \cdot 5$. The two numbers have no primes in common, but the values of the exponents are 3, 2 and 1 in the case of 954569, and 2, 3, and 1 in the case of 540. The number of divisors of 954569 is $3 \cdot 4 \cdot 2 = 24$ is the same as the number of divisors of 540. Thus, *it is the set of values of the exponents $\{2, 3, 1\}$ on the primes in the factorization that determines the number of divisors, not the size of the number or the set of primes themselves.*

✻

Challenge Problem. Tabulate the divisors of 2100.

✻

We close this section with a summary of the facts on the form and number of positive divisors of a positive integer n, and with an interesting application of these facts needed for our discussion of The Locker Problem in the next section.

OBSERVATION 1.1. *If*

$$n = p_1^{a_1} p_2^{a_2} \cdots p_t^{a_t},$$

is the prime power factorization of n, then an integer is a positive divisor of n precisely when it has the form

$$p_1^{b_1} p_2^{b_2} \cdots p_t^{b_t},$$

where $0 \le b_1 \le a_1$, $0 \le b_2 \le a_2$, ..., $0 \le b_t \le a_t$. Moreover, there are

$$(a_1 + 1)(a_2 + 1)(a_3 + 1) \cdots (a_t + 1)$$

positive divisors of n.

We examine the integer $N = (a_1 + 1)(a_2 + 1)(a_3 + 1) \cdots (a_t + 1)$ more closely. The integer N is even if any one (or more) of its factors is even. We conclude, therefore, that the number N of divisors of an integer n is even precisely when *any one* of the integers a_1, a_2, ...,a_t is odd. On the other hand, we see that the number of divisors of an integer n is odd exactly when *every* integer a_1, a_2, ...,a_t is even.

Recall that a positive integer is called a *perfect square* if it is the square of another positive integer. For example, $4 = 2^2$, $36 = 6^2 = 2^2 \cdot 3^2$, $169 = 13^2$ and $784 = (28)^2 = (2^2)^2 \cdot 7^2$ are perfect squares.

Let us examine the prime power factorization of a perfect square. We start with an example $144 = 12^2$. The prime power factorization of 144 is

$$144 = 2^4 \cdot 3^2.$$

We see that 2 and 3 are the primes in the factorization and, in the factorization, the exponent of each of these primes is an even number. Thus, each prime power factor of 144 is a perfect square. On the other hand, suppose the prime power factorization of an integer is a product of primes with even exponents, such as

$$3600 = 2^4 \cdot 3^2 \cdot 5^2.$$

When we write
$$3600 = 2^4 \cdot 3^2 \cdot 5^2 = (2^2 \cdot 3 \cdot 5)^2$$
it becomes clear that the integer $3600 = (2^2 \cdot 3 \cdot 5)^2 = 60^2$ is a perfect square.

What does this imply about the number N of positive divisors of 3600? Since every exponent in the prime power factorization is even, the number $N = (4+1)(2+1)(2+1) = 45$ is clearly odd.

The same is true more generally. If
$$n = p_1^{a_1} p_2^{a_2} \cdots p_t^{a_t},$$
is the prime power factorization of the integer $n > 1$, then n is a perfect square precisely when each of the exponents $a_1, \ldots, a_t$ is even. For, if that is the case, then $a_1 = 2b_1, \ldots, a_t = 2b_t$, for some positive integers $b_1, \ldots, b_t$. Consequently,
$$n = p_1^{2b_1} p_2^{2b_2} \cdots p_t^{2b_t} = (p_1^{b_1} p_2^{b_2} \cdots p_t^{b_t})^2,$$
and n is a perfect square. Moreover, we see that if every exponent a_i is even then the number $N = (a_1 + 1)(a_2 + 1)(a_3 + 1) \cdots (a_t + 1)$ of positive divisors of n is odd.

We summarize this discussion in the following obervation.

OBSERVATION 1.2. *Let*
$$n = p_1^{a_1} p_2^{a_2} \cdots p_t^{a_t},$$
be a positive integer. The number $N = (a_1+1)(a_2+1)\cdots(a_t+1)$ of divisors of n is even precisely when any one, or more, of the integers a_1, a_2, $\ldots, a_t$ is odd. The number N is odd exactly when every integer a_1, a_2, $\ldots, a_t$ is even.

Consequently, n is a perfect square precisely when every exponent $a_1, \ldots, a_t$ is even and this occurs precisely when the number $N = (a_1 + 1)(a_2 + 1) \cdots (a_t + 1)$ of its positive divisors is odd.

The final statement of the observation confirms what we already know, namely, "Divisors come in pairs except for perfect squares."

2. The Locker Problem

The Problem. There are K lockers numbered consecutively from 1 to K lining a hallway. All locker doors are closed. Student 1 walks down the hallway and opens all of the locker doors. Student 2 closes the doors of all lockers with numbers divisible by 2, that is, with numbers 2, 4, 6, 8, ... Student 3 reverses the position of the doors of all lockers with numbers divisible by 3, that is, with numbers 3, 6, 9, 12, ... Student 4 reverses the position of the doors of all lockers with numbers divisible by 4, that is, with numbers 4, 8, 12, 16, ... At this point, locker 1 is open, locker 2 is closed, locker 3 is closed, locker 4 is open, locker 5 is open, locker 6 is open, locker 7 is open, locker 8 is open, ... Below, we tabulate the positions of the locker doors after Students 1 - 4 have walked down the hall. We place an entry in a column of the table only when the position of the door of that locker changes.

	#1	#2	#3	#4	#5	#6	#7	#8	...
	closed	closed	closed	closed	closed	closed	closed	closed	...
1	open	open	open	open	open	open	open	open	...
2		closed		closed		closed		closed	...
3			closed			open			...
4				open				open	...

This pattern continues: Student h reverses the position of the doors of all lockers with numbers divisible by h. Finally, Student K reverses the position of the door of locker K. The question is: Which locker doors are open after Student K has reversed the position of the door of locker K?

Discussion. There are many entries on the Internet devoted to the Locker Problem. They run the gamut from applets to videos. You might enjoy looking at some of them. Unfortunately, many of the accompanying explanations of the problem are unsatisfactory.

We will focus our discussion on what exactly is going on in the Locker Problem, and how to solve it. It is easy to be overwhelmed by all the lockers. A key to understanding the problem is to study what happens to the door of just one locker.

To start, we focus on one locker, say locker $j > 1$. If Student h walks by locker j and h is not a divisor of j, then there is no change in the position of the door of locker j. The door of locker j changes position only if Student h walks by where h is a divisor of j. Since $h \mid j$ implies that $h \leq j$, the final time that the door of locker j is reversed is when $h = j$. It follows that **the number of changes of the position of the door of a locker is equal to the number of divisors of the locker number.**

Example. For an explicit example, we take $K = 16$ and study specific lockers. We look at locker 5 first. The integer 5 is prime; it has exactly two divisors 1 and 5. So the door of locker 5 is opened by Student 1 and closed by Student 5. The integer 5 has an even number of divisors so the door of locker 5 changes position an even number of times. Consequently, the locker door reverts back to its closed position.

Since $6 = 3 \cdot 2$, the integer 6 has four divisors: 1, 2, 3 and 6. The door to locker 6 is opened by Student 1, closed by Student 2, opened by Student 3 and closed by Student 6. Again, we have an example of a locker number with an even number of divisors, so the locker door is closed after all students have walked down the hallway.

Next, we look at 15. The integer $15 = 3 \cdot 5$ has four divisors: 1, 3, 5 and 15. The door of locker 15 is opened by Student 1, closed by Student 3, opened by Student 5 and shut by Student 15. Again, we see that the locker door will revert to its closed position exactly when the locker number has an even number of divisors.

All integers from 1 through 16 except 1, 4, 9 and 16 have an even number of divisors. The integers 1, 4, 9 and 16 are perfect squares, and perfect squares are, as we know, the only positive integers that have an odd number of divisors. Consequently, a locker door with a number that is a perfect square, say s^2, will be opened by Student s^2 and remain open after all students have walked down the hall. In particular, the door of locker 1 is opened by Student 1 and it stays open. The integer 4 has three divisors: 1, 2 and 4. The door of locker 4 is opened by Student 1, closed by Student 2 and opened by Student 4. The number $9 = 3^2$ has three divisors: 1, 3 and 9. The position of the door of locker 9 changes exactly 3 times when Students 1, 3 and 9 walk down the hallway. Student 1 opens the door, Student 3 closes it, Student 9 opens it, and the door remains open after all students have walked down the hallway. The number $16 = 4^2$ has five divisors: 1, 2, 4, 8 and 16. It is clear that after Students 1, 2, 4, and 8 have walked down the hallway, Student 16 will find the door of locker 16 closed, and Student 16 will open it. So it is the doors of the lockers with numbers that are perfect squares that are open after all students have walked down the hall.

The same reasoning applies to the general case where there are K lockers. We have the solution to the Locker Problem: **The lockers with open doors are precisely the lockers with numbers that are perfect squares.**

3. The Greatest Common Divisor, the Least Common Multiple and the Least Common Denominator

In this section, we show how factorization into primes can be used to calculate the greatest common divisor of two positive integers a and b. We also define the companion notion of least common multiple of a and b.

Recall, from Seminar 3, that d is the greatest common divisor of positive integers a and b if $d \mid a$ and $d \mid b$, and if every positive integer dividing both a and b is less than or equal to d. In other words, d is the largest integer that divides both a and b. In this seminar, we look again at $d = \gcd(a, b)$ and, for the first time, at a complementary concept.

A positive integer m is the *least common multiple* of two positive integers a and b if $a \mid m$ and $b \mid m$, and if every positive integer multiple of a and b is greater than or equal to m. In other words, m is the smallest positive integer that is divisible by both a and b. We denote the least common multiple m of a and b by $m = \text{lcm}[a, b]$.

Observe that the least (or lowest) common denominator of two fractions c/d and e/f is the least common multiple of their denominators: $\mathrm{lcm}[d, f]$. How do we calculate $\gcd(a, b)$ and $\mathrm{lcm}[a, b]$? For small a and b, it is not difficult to calculate $\gcd(a, b)$ by examining divisors. Nor is it difficult to calculate $\mathrm{lcm}[a, b]$ by writing down the multiples of a and the multiples of b and picking out the least one that is common to both.

For example, the divisors of 15 are 1, 3, 5 and 15, while the divisors of 21 are 1, 3, 7 and 21. Consequently, $\gcd(15, 21) = 3$. Since the multiples of 15 are 15, 30, 45, 60, 75, 90, 105,..., and the multiples of 21 are 21, 42, 63, 84, 105, ..., it follows that $\mathrm{lcm}[15, 21] = 105$. However, as we saw in Seminar 3, the Euclidean Algorithm is, far and away, the best method to compute the greatest common divisor. In fact, we will see later that the Euclidean Algorithm can also be used to calculate the least common multiple.

In this section, we introduce you to a method for computing $\gcd(a, b)$ and $\mathrm{lcm}[a, b]$ from the prime factorizations of a and b. We must emphasize the point, mentioned earlier, that the computation of prime factorizations of large numbers is frequently quite difficult and explains why the Euclidean Algorithm is the preferred method for computing the gcd. We will also see that, once $\gcd(a, b)$ has been calculated, the computation of $\mathrm{lcm}[a, b]$ is no trouble at all.

We will use examples to describe how to compute the greatest common divisor and the least common multiple by means of prime factorizations.

To begin each calculation, we write down the prime factorizations of the numbers a and b. If the factorizations of a and b have no primes in common, then, as we discussed in Section 1, no divisor of a greater than 1 can divide b, and no divisor of b greater than 1 can divide a. Consequently, we have $\gcd(a, b) = 1$. We will deduce what $\mathrm{lcm}[a, b]$ must be in this case momentarily. An example of this case is $a = 21 = 3 \cdot 7$ and $b = 22 = 2 \cdot 11$. We have $\gcd(21, 22) = 1$.

Consider the example where $a = 15$ and $b = 21$. We show how to calculate $\gcd(15, 21)$ and $\mathrm{lcm}[15, 21]$ using the prime factorization of 15 and of 21. We have

$$15 = 3 \cdot 5; \quad 21 = 3 \cdot 7.$$

We see that the prime 3 appears in both factorizations above, the prime 5 appears in one factorization and the prime 7 appears in the other. Since 3 is the only prime common to both factorizations, it follows that $\gcd(15, 21) = 3$.

The least common multiple of 15 and 21 has to be a multiple of both 15 and 21, and the smallest such. Consequently, $\mathrm{lcm}[15, 21]$ is the product of all of the primes appearing in one factorization or the other with the exponent of each such prime equal to the *maximum* number of times it appears in one factorization or the other. Thus, $\text{lcm}[15, 21] = 3 \cdot 5 \cdot 7 = 105$.

Next, we look at an example with numbers having some repeated primes in their prime factorizations. We compute $\gcd(84, 90)$ and $\text{lcm}[84, 90]$. The first step is to factor both numbers into primes.

$$84 = 2^2 \cdot 3 \cdot 7; \quad 90 = 2 \cdot 3^2 \cdot 5.$$

To compute $\gcd(84, 90)$, we write down the distinct prime numbers that are common to both factorizations. In this case, there are two, namely, 2 and 3. In the factorization of 84, the prime 2 has exponent 2 and in the factorization of 90, the prime 2 has exponent 1. In the factorization of 84, the prime 3 has exponent 1 and in the factorization of 90, the prime 3 has exponent 2. Since $\gcd(84, 90)$ divides both 84 and 90, their greatest common divisor is the product of all the common primes with the proviso that the common prime has exponent equal to the *minimum* of the exponents it has in one factorization or the other. So, for 84 and 90, the greatest common divisor is the product $2 \cdot 3 = 6$. Observe that if the exponent of 2 or 3 were any higher in the product, it would not divide both 84 and 90.

Since $\text{lcm}[84, 90]$ is a multiple of 84 and a multiple of 90, the least common multiple of 84 and 90 must be a product of all the distinct prime factors appearing in one factorization or the other, each having exponent equal to the *maximum* of the exponents that prime has in one factorization or the other. Consequently, $\text{lcm}[84, 90] = 2^2 \cdot 3^2 \cdot 5 \cdot 7 = 1260$.

∗

Seminar Questions.
(i) When, if ever, is $\text{lcm}[a, b]$ is equal to the product ab?
(ii) Suppose $a \mid b$. Calculate $\gcd(a, b)$ and $\text{lcm}[a, b]$.

∗

Comments on the Seminar Question.
(i) If $\gcd(a, b) = 1$, then a and b have no primes in common. In this case $\text{lcm}[a, b]$ is the product of all the primes appearing in one factorization or the other, each having exponent equal to the exponent of that prime in the factorization in which it appears. Consequently, when $\gcd(a, b) = 1$, then (and only then) $\text{lcm}[a, b]$ is equal to the product ab. We return now to the example $a = 21$ and $b = 22$. We found that $\gcd(21, 22) = 1$. It follows that $\text{lcm}[21, 22] = 21 \cdot 22 = 462$.
(ii) If $a \mid b$, then $\gcd(a, b) = a$, and $\text{lcm}[a, b] = b$.

∗

Seminar Exercise. Using prime factorizations, compute

(i) $\gcd(32, 100)$, and $\text{lcm}[32, 100]$; and
(ii) $\gcd(132, 180)$, and $\text{lcm}[132, 180]$.

(iii) In both cases above, calculate $\gcd(a, b) \cdot \operatorname{lcm}[a, b]$. What are these numbers? Does this suggest a method for computing the $\operatorname{lcm}[a, b]$ from the $\gcd(a, b)$ or vice versa?

∗

Comments on the Seminar Exercise.

(i) We have

$$32 = 2^5; \quad 100 = 2^2 \cdot 5^2.$$

These factorizations have only the prime 2 in common and 2 is squared in the prime power factorization of 100. Thus, $\gcd(32, 100) = 2^2$. To calculate $\operatorname{lcm}[32, 100]$, we observe that 2 appears to the fifth power in the prime power factorization of 32, and 5 is squared in the prime factorization of 100. It follows that $\operatorname{lcm}[32, 100] = 2^5 \cdot 5^2 = 32 \cdot 25 = 800$.

(ii) The prime factorizations are

$$132 = 2^2 \cdot 3 \cdot 11; \quad 180 = 2^2 \cdot 3^2 \cdot 5.$$

The distinct common primes are 2 and 3. The prime 2 has exponent 2 in both factorizations. The prime 3 has exponent 1 in the factorization of 132 and 2 in the factorization of 180. It follows that $\gcd(132, 180)$ is a product of 2 with exponent 2 and 3 with exponent 1, that is, $\gcd(132, 180) = 2^2 \cdot 3 = 12$. As for $\operatorname{lcm}[132, 180]$, we have 2^2 and 3^2 are factors as well as the primes 11 and 5, each with exponent 1. Thus, $lcm[132, 180] = 2^2 \cdot 3^2 \cdot 5 \cdot 11 = 1980$.

(iii) We have

$$\gcd(32, 100) \cdot \operatorname{lcm}[32, 100] = 4 \cdot 800 = 3200.$$

and

$$\gcd(132, 180) \cdot \operatorname{lcm}[132, 180] = 12 \cdot 1980 = 23760.$$

Thus, in both of these examples, $\gcd(a, b) \cdot \operatorname{lcm}[a, b] = ab$.

Note. Before we summarize our findings, we note that if m and n are two positive integers then the product of the minimum, or the smaller, of the two integers and the maximum, or the larger, of the two integers is the product mn of the two integers. Thus, in symbols, we have the minmax rule:

$$\min(m, n) \max(m, n) = mn.$$

Summary. By definition, $\gcd(a, b)$ is the largest positive integer that divides both a and b. As we have seen with the examples above, it follows that $\gcd(a, b)$ is the product of all the primes common to the prime power factorizations of a and b, with the proviso that each common prime has exponent equal to the *minimum* of the exponents of that prime in one factorization or the other. Whereas, $\operatorname{lcm}[a, b]$ is the smallest positive integer that is a multiple of both a and b. It follows that $\operatorname{lcm}[a, b]$ is the product of all of the primes appearing in one factorization or the other with the exponent of each such prime equal to the *maximum* number of times it appears in one factorization or the other. By applying the minmax rule above to the exponents, we have

$$\gcd(a,b) \cdot \operatorname{lcm}[a,b] = ab.$$

Consequently, the least common multiple can be computed from the greatest common divisor and vice-versa.

$$\operatorname{lcm}[a,b] = \frac{ab}{\gcd(a,b)}; \quad \gcd(a,b) = \frac{ab}{\operatorname{lcm}[a,b]}.$$

The least common multiple plays an important role in addition of fractions, for the least, or lowest, common denominator of two fractions is the least common multiple of the denominators of the fractions. Consequently, the equation

$$\operatorname{lcm}[a,b] = \frac{ab}{\gcd(a,b)}$$

leads to a method for finding the least common denominator of two fractions by means of the Euclidean Algorithm! We studied the Euclidean Algorithm in Seminar 3, Section 5. You may want to review it before reading further here.

We illustrate this method of finding the least common denominator with a simple example first. To compute the sum

$$\frac{13}{21} + \frac{5}{24},$$

we find the least (or lowest) common denominator of 21 and 24. We use the Euclidean Algorithm to calculate $\gcd(21,24)$.

$$24 = 21 \cdot 1 + 3$$
$$21 = 3 \cdot 7 + 0$$

Thus, $\gcd(21,24) = 3$. Next, we use the equation for the lcm.

$$\operatorname{lcm}[21,24] = \frac{21 \cdot 24}{\gcd(21,24)} = \frac{504}{3} = 168.$$

To add the fractions, note that $168 = 21 \cdot 8$ and $168 = 24 \cdot 7$. Consequently,

$$\frac{13}{21} = \frac{13 \cdot 8}{168} = \frac{104}{168}$$

and

$$\frac{5}{24} = \frac{5 \cdot 7}{168} = \frac{35}{168}.$$

The sum of the fractions is

$$\frac{13}{21} + \frac{5}{24} = \frac{104}{168} + \frac{35}{168} = \frac{139}{168}.$$

Of course, you know that to add these fractions or any other fractions, it is sufficient to find *any* common multiple of the denominators. For example, you can always use the product and then reduce to lowest terms, if necessary. (Even with the least common denominator, the sum may not be in lowest terms, as we shall see.) We use the product of the denominators as the common denominator in the following example.

$$\frac{13}{21} + \frac{5}{24} = \frac{13 \cdot 24}{21 \cdot 24} + \frac{5 \cdot 21}{21 \cdot 24}$$
$$= \frac{312}{504} + \frac{105}{504} = \frac{417}{504}$$
$$= \frac{\cancel{3} \cdot 139}{\cancel{3} \cdot 168} = \frac{139}{168}.$$

In Seminars 8, 9 and 10, we discuss in detail the arithmetic of fractions including addition of fractions.

✳

Seminar Exercise.   Use the Euclidean Algorithm to find the least (or lowest) common denominator and add the following fractions:
(i)

$$\frac{7}{78} + \frac{5}{84}$$

(ii)

$$\frac{53}{280} + \frac{41}{343}$$

✳

Comments on the Seminar Exercise.
(i) By the Euclidean Algorithm,

$$84 = 78 \cdot 1 + 6$$
$$78 = 6 \cdot 13 + 0.$$

Thus, $\gcd(78, 84) = 6$, and

$$\mathrm{lcm}[78, 84] = \frac{78 \cdot 84}{6} = 1092.$$

We have, therefore,

$$\frac{7}{78} + \frac{5}{84} = \frac{7 \cdot 14}{1092} + \frac{5 \cdot 13}{1092} = \frac{98}{1092} + \frac{65}{1092} = \frac{163}{1092}$$

(ii) The numbers in this problem are large. Just do it step by step. By the Euclidean algorithm,

$$343 = 280 \cdot 1 + 63$$
$$280 = 63 \cdot 4 + 28$$
$$63 = 28 \cdot 2 + 7$$
$$28 = 7 \cdot 4 + 0.$$

Thus, $\gcd(280, 343) = 7$, and

$$\text{lcm}[280, 343] = \frac{280 \cdot 343}{7} = \frac{96040}{7} = 13720.$$

Consequently,

$$\frac{53}{280} + \frac{41}{343} = \frac{53 \cdot 49}{13720} + \frac{41 \cdot 40}{13720} = \frac{2597}{13720} + \frac{1640}{13720} = \frac{4237}{13720}.$$

4. Secret Codes, a Game of Integer Divisors, II

Using the same code as in Seminar 2, Section 4, we will use our knowledge of prime numbers, the Fundamental Theorem of Arithmetic and finding divisors to decode some secret messages.

A	B	C	D	E	F	G	H	I	J	K	L	M	N	O	P	Q
1	2	3	4	5	6	7	8	9	10	11	12	13	14	15	16	17

R	S	T	U	V	W	X	Y	Z
18	19	20	21	22	23	24	25	26

In Seminar 2, we decoded single words. This time, the idea of the game is to decode simple messages. Each code number represents a word in the message. For example, consider the code consisting of the two numbers

$$105 \quad 575$$

We decode this message one word at a time.

Consider 105. First, we find its prime power factorization. 105 is clearly divisible by 5, and $105 = 5 \cdot 21$. Consequently, the prime power factorization of 105 is

$$105 = 3 \cdot 5 \cdot 7;$$

with distinct primes 3, 5, and 7, each with exponent 1. The next step is to find the code letters of the positive divisors of 105 *that are less than or equal to* 26. Using the methods of the previous section, we find that these divisors are: 1, 3, 5, 7, 15 and 21. The corresponding code letters are A, C, E, G, O, and U. Recall that the code numbers of words are obtained by taking the product of the code numbers of the letters in the word. We must use the letters A, C, E, G, O, and U to make a word with code number equal to 105. Observe that since all the exponents in the prime power factorization are equal to 1, no letter, other than A, may be used more than once. One word that works is GO since $7 \cdot 15 = 105$. Other possible code words are AGO and CAGE.

We use the same analysis on the code number 575.

$$575 = 5^2 \cdot 23.$$

The divisors of $575 \leq 26$ are 1, 5, 23 and 25. The corresponding code letters are A, E, W, and Y. These letters form the words EWE and AWAY, each with code number 575. The word AWAY combines with GO, to form the message

GO AWAY.

Thus, despite the defect that our code does not associate unique words to code numbers, we do obtain a code message. We will never know if this is the one the encoder intended. As you might suspect, this code is used primarily for educational purposes and for games.

Summary. Here are the steps to decoding a code number in a message.

  (i) Find the prime power factorization of the code number of the word.
 (ii) Find all of the divisors of the code number that are less than or equal to 26.
(iii) For each number in (ii), write down the corresponding code letter.
 (iv) Use these letters to make a word whose corresponding code number, obtained by taking the product of the code numbers of the letters in the word, is equal to the given code number. Remember that you may use as many A's as you please.

Finally, after you have decoded all the numbers, make sure your words form a message that makes sense.

✳

Seminar Code Game 1. Divide your class into small groups of students working with one another to find a secret message.

  (i) The first message is:

700 208.

Follow the steps outlined above to decode it.
 (ii) Decode this message given the following code numbers:

6500 20 146300.

(iii) Here is an encoded message

105 1800 1 9900.

Decode a four word message.

✳

Comments on Seminar Code Game 1.

(i) $700 = 2^2 \cdot 5^2 \cdot 7$. The divisors of 700 less than or equal to 26 are 1, 2, $2^2 = 4$, 5, 7, $2 \cdot 5 = 10$, $2 \cdot 7 = 14$, $2^2 \cdot 5 = 20$, and $5^2 = 25$. The corresponding code letters are A, B, D, E, G, J, N, T, Y. One word with code number 700 is GET. You may have found others.

$208 = 13 \cdot 2^4$. The divisors of 208 less than or equal to 26 are 1, 2, $2^2 = 4$, $2^3 = 8$, 13, $2^4 = 16$, and $2 \cdot 13 = 26$. The corresponding code letters are A, B, D, H, M, P and Z. One word with code number 208 is MAP. So one message (that makes sense) is

GET MAP

(ii) $6500 = 2^2 \cdot 5^3 \cdot 13$. The divisors of 6500 less than or equal to 26 are 1, 2, $2^2 = 4$, 5, $2 \cdot 5 = 10$, 13, $2^2 \cdot 5 = 20$ and $5^2 = 25$. The corresponding code letters are A, B, D, E, J, M, T and Y. One word with code number 6500 is MEET. (Note that we can to use two E's or one Y because of the factor 5^2.)

$20 = 2^2 \cdot 5$. The divisors of 20 less than or equal to 26 are 1, 2, $2^2 = 4$, 5, $2 \cdot 5 = 10$ and 20. The corresponding code letters are A, B, D, E, J, T. One word with code number 20 is AT which combines with MEET to form MEET AT, a good start.

$146300 = 2^2 \cdot 5^2 \cdot 7 \cdot 11 \cdot 19$. The divisors of 146300 less than or equal to 26 are 1, 2, $2^2 = 4$, 5, 7, $2 \cdot 5 = 10$, 11, $2 \cdot 7 = 14$, 19, $2^2 \cdot 5 = 20$, $2 \cdot 11 = 22$ and $5^2 = 25$. The corresponding code letters are A, B, D, E, G, J, K, N, S, T, V and Y. One word with code number 146300 is SEVEN. We obtain the message:

MEET AT SEVEN

(iii) The divisors and associated letters of 105 were found earlier. We deciphered three possible words: GO, AGO and CAGE. There probably are not very many more, since the number of divisors of 105 is relatively small.

We have $1800 = 2^3 \cdot 3^2 \cdot 5^2$. The divisors of 1800 less than or equal to 26 are: 1, 2, 3, $2^2 = 4$, 5, 6, $2^3 = 8$, $3^2 = 9$, $2 \cdot 5 = 10$, $3 \cdot 4 = 12$, $3 \cdot 5 = 15$, $2 \cdot 3^2 = 18$, $2^2 \cdot 5 = 20$, $2^3 \cdot 3 = 24$ and $5^2 = 25$. The corresponding letters are A, B, C, D, E, F, H, I, J, L, O, R, T, X and Y. Since 1800 has many divisors, there are undoubtedly many words with corresponding code number equal to 1800. One word that works is FLY. We can combine this with GO to obtain GO FLY. The third code number obviously corresponds to A, so we have, thus far, GO FLY A. Let's see if we can cut short all the testing of the code number 9900 by checking if the word KITE has code number 9900. Indeed it has, so we have the message

GO FLY A KITE

✳

Seminar Code Game 2. Divide your class into small groups of students working with one another.

(i) Ask each group to encode a message.

(ii) Ask each group to pass its encoded message to another group to decode.

✳

Seminar Question. What other numerical information could you add to the encoded words to lower the possibility of decoding it into more than one message?

✳

Comments on the Seminar Question. You might give, for example, the sum of the code numbers of the letters in each word following the code number of the word. If we do this for the message GO AWAY, the encoded message becomes

105   22   575   50.

Seminar 6

Modular Arithmetic and Divisibility Tests

1. Preliminaries

Before we begin the subject matter of the seminar, we take a moment to recall that all integers, not only the positive ones, can be classified as either even or odd. An integer is *even* if it is divisible by 2 and can, therefore, be written in the form $2n$, for some integer n. The integers -10, 0 and 6 are examples of even integers. An integer is *odd* if it is not divisible by 2. In fact, every odd integer is of the form $2n+1$, for some integer n. The integers 5, -7 and -21 are odd integers.

✳

Seminar Exercise. Show that each of the odd integers 5, 7, -7 and -21 can be written in the form $2n + 1$, for some integer n, positive, negative or zero.

✳

Comments on the Seminar Exercise. Observe that $7 = 2(3) + 1$, and $-7 = 2(-4) + 1$.

2. Introduction

We are most familiar with number systems, such as the integers and the rational numbers, where there is no "first" and also no "last" number. In this seminar, we introduce you to the concept of congruence and to number systems that have only a finite number of elements. There are many instances in daily life where these number systems come into play. We begin with two such examples.

3. Examples of Congruence

3.1. Congruence Modulo 2

For the first example, consider the operation of an ordinary light switch. There are two possible ways we interact with a light switch: do nothing and flip the switch. If the light is OFF and we do nothing, the light remains OFF.

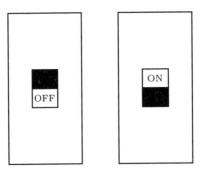

If we flip the switch, the light turns ON. If we flip the switch a second time, the light turns OFF, the same as if we did nothing. If we flip the switch three times, the light is ON, four times, the light is OFF. No matter how many times we flip the switch, there are only two possible results: the light is OFF or the light is ON. If the switch starts in the OFF position, and if the number of flips of the switch is even, the light stays OFF, if odd, the light is ON.

How does this relate to number theory? Start with the light OFF and let 0 correspond to "doing nothing" to the switch and let 1 correspond to one flip of the switch. Each additional flip of the switch adds 1 to the previous number of flips. Thus, repetition of an action, that is, of either doing nothing or flipping the switch, corresponds to addition. Consequently, all the nonnegative even numbers give the same result as 0, that is, doing nothing, and all the nonnegative odd numbers give the same result as 1, one flip of the switch. Every even positive integer has remainder 0 when we divide it by 2 and every odd positive integer has remainder 1 when we divide it by 2. We express this mathematically by saying that every even nonnegative integer, $0, 2, 4, 6, 8, 10, \ldots$, is *congruent to* 0 *modulo* 2, and every odd positive integer, $1, 3, 5, 7, 9, 11, \ldots$, is *congruent to* 1 *modulo* 2. We use the symbol $\equiv$ to stand for *is congruent to* and write, for example, $2 \equiv 0$ (mod 2), $10 \equiv 0$ (mod 2), $3 \equiv 1$ (mod 2) and $25 \equiv 1$ (mod 2).

✳

Seminar Exercise. Tell whether each of the following numbers is congruent to 0 (mod 2) or is congruent to 1 (mod 2). Write your answers in the form $a \equiv 0$ (mod 2), or $a \equiv 1$ (mod 2).

3, 23, 55552, 624005389, 1

✳

Comments on the Seminar Exercise. We have $3 \equiv 1$ (mod 2), $23 \equiv 1$ (mod 2), $55552 \equiv 0$ (mod 2), $624005389 \equiv 1$ (mod 2) and $1 \equiv 1$ (mod 2).

✳

Before we give the definition of congruence in general, we consider one more example.

3.2. Congruence Modulo 12

Next let us look at a clock. Our clock has only one hand, the hour hand. If it is 11 o'clock now and we have an appointment in 3 hours, we know that means that the appointment is at 2 o'clock. The hour hand moves around the clock from 11 to 12 and then starts around again to 1 and finally to 2. When the hand passes 12, 12 becomes 0 and we have $11 + 3 = 11 + 1 + 2 = 12 + 2 = 0 + 2 = 2$. So, on the clock, 14 is represented by 2.

We replace the 12 on the face of the clock by 0, and, unless otherwise stated, we consider 0 to be the starting point of the hour hand. For any number of integer hours the hour hand points to one of the twelve numbers $0, \ldots, 11$ on the face of the clock. For example, at 28 hours, the hand starts at 0, passes 0 two times and points to 4. At 52 hours, the hand starts at 0, passes 0 four times and points to 4.

Observe that *the hour hand points to the same hour on the clock for all pairs of integers whose difference is a multiple of* 12. For the examples above, $28 - 4 = 24 = 2 \cdot 12$, and $52 - 4 = 48 = 4 \cdot 12$. Another way to write each of these equations is: $28 = 2 \cdot 12 + 4$, and $52 = 4 \cdot 12 + 4$. Thus, *the hour*

hand points to the hour on the clock equal to the remainder upon division by 12, and we say that all numbers with the same remainder upon division by 12 are congruent modulo 12. In the example of the ON-OFF switch, we have congruence modulo 2; in this example, we have congruence modulo 12.

Negative integers do not make much sense in the switch example. For the clock, we interpret negative integers as counterclockwise rotation of the hand. The hand moves clockwise for positive hours and counterclockwise for negative hours. We remind you that the division algorithm can be extended to allow the dividend a to be negative. (See the end of Section 3 in Seminar 3.) The divisor b is always assumed to be positive and the restriction on the remainder r, $0 \leq r < b$ remains the same. Consequently, for *any* integer a, we may use the division algorithm to divide a by 12 and determine the remainder r :

$$a = 12 \cdot q + r,$$

where $0 \leq r < 12$. Note that if $a < 0$, then $q < 0$, but r is always nonnegative. Consequently, for a positive, zero or negative, r is one of the integers 0, 1, 2, 3, 4, 5, 6, 7, 8, 9, 10, 11, and a and r are congruent modulo 12. We write

$$a \equiv r \pmod{12}.$$

For example, $62 = 12 \cdot 5 + 2$, so $62 \equiv 2 \pmod{12}$, For an example with negative dividend, take -61. By the extended division algorithm, $-61 = 12 \cdot (-6) + 11$ so $-61 \equiv 11 \pmod{12}$.

✳

Seminar Exercise.
(i) For each of the following integers a, find an integer r with $0 \leq r < 12$, so that $a \equiv r \pmod{12}$.

13, 64, 323, 509, 5280, 286, −55.

(ii) Match each integer in the left hand column to one in the right hand column that is congruent to it mod 12. More than one of the integers in the left hand column may match with the same integer in the right hand column.

116	2
205	11
62	5
321	10
509	8
108	1
673	0

✳

Comments on the Seminar Exercise.

(i) The integer r is the remainder when we divide each of the given integers a by 12 using the division algorithm. $13 = 12 \cdot 1 + 1$, so $13 \equiv 1 \pmod{12}$. $64 = 12 \cdot 5 + 4$, so $64 \equiv 4 \pmod{12}$.
$323 = 12 \cdot 26 + 11$, so $323 \equiv 11 \pmod{12}$.
$509 = 12 \cdot 42 + 5$, so $509 \equiv 5 \pmod{12}$.
$5280 = 12 \cdot 440 + 0$, so $5280 \equiv 0 \pmod{12}$.
$286 = 12 \cdot 23 + 10$, so $286 \equiv 10 \pmod{12}$.
$-55 = 12 \cdot (-5) + 5$, so $-55 \equiv 5 \pmod{12}$.

(ii) For the first number in the left hand column, we have $116 \equiv 8 \pmod{12}$. Since the number 205 and the number 673 have the remainder 1 upon division by 12, $205 \equiv 1 \pmod{12}$ and $673 \equiv 1 \pmod{12}$.
For the remaining numbers we have $62 \equiv 2 \pmod{12}$, $321 \equiv 9 \pmod{12}$, $509 \equiv 5 \pmod{12}$ and $108 \equiv 0 \pmod{12}$.

✳

We are ready to define congruence modulo any integer greater than 1.

4. Congruence

We introduce a relationship between integers based on divisibility. Let m be an integer greater than 1. Let a and b be integers. We say that say that a *is congruent to* b *modulo* m, if $a - b$ is divisible by m, or, what is the same thing, if $a - b$ is multiple of m. We write

$$a \equiv b \pmod{m},$$

and say "a is congruent to b mod m." The number m is called the *modulus*.

For example, 4 is congruent to 0 (mod 2) because 2 divides $4 - 0 = 4$. Moreover, any multiple of 2 is congruent to 0 (mod 2), because $2k - 0 = 2k$ is a multiple of 2. The integer 15 is congruent to 1 (mod 7), because $15 - 1 = 14 = 2 \cdot 7$ is a multiple of 7. Integers of the form $15 + 7k$, for any integer k, are also congruent to 1 (mod 7), because $15 + 7k - 1 = 14 + 7k = 7(2 + k)$ is a multiple of 7. As another example, we have -1 is congruent to 2 (mod 3), because 3 divides $-1 - 2 = -3$. Then again, integers of the form $-1 + 3k$, for any integer k are also congruent to 2 (mod 3).

It follows from the definition that $a \equiv 0 \pmod{m}$ precisely when m divides a.

Next, we highlight a nice way to think about congruence. Let m be any integer greater than 1. For each integer a, one of the integers $0, 1, \ldots, m-2$, $m - 1$ is congruent to a (mod m). For if we apply the division algorithm to divide a by m, we have

$$a = qm + r,$$

where $0 \le r < m$. Thus, $a - r = qm$, so $a \equiv r \pmod{m}$.
(Of course, there are also many other integers congruent to a (mod m).

What we have just shown you is a way to find an integer r congruent to a (mod m) with the property that $0 \leq r < m$.

For example, if we have $m = 6$, and take $a = 23$, then one of the integers 0, 1, 2, 3, 4, or 5 is congruent to 23 (mod 6). Which one is it? It is the remainder after division by 6. To find that remainder, we divide 23 by 6 :

$$23 = 6 \cdot 3 + 5,$$

and so,

$$23 - 6 \cdot 3 = 5.$$

By the definition of congruence, $23 \equiv 5$ (mod 6). Thus, you already have the tools to do these problems. We believe you will be interested in their applications.

If the modulus is 2, then the possible remainders upon division of any integer by 2 are 0 and 1. For example, $-10, -4, 0, 4, 18$ and all even numbers are congruent to 0 mod 2, and $-13, -1, 3, 7, 31$ and all odd numbers are congruent to 1 mod 2,

You have two possible ways to determine whether two integers a and b are congruent (mod m). *They are:*
(i) *Use the definition of congruence and decide whether $a - b$ is divisible by m, or not.*
(ii) *Apply the division algorithm to divide both a and b by m, and compare the remainders. The remainders are equal precisely when $a \equiv b$ (mod m).)*

✳

Seminar Exercise. Label each of the following statements as True (T) or False (F). Justify your answer.
  (i) $117 \equiv 0$ (mod 9).
 (ii) $26 \equiv 1$ (mod 5).
(iii) $-8 \equiv -3$ (mod 11)
 (iv) $-49 \equiv -7$ (mod 6)
  (v) $5 \cdot 3 \equiv 1$ (mod 2)
 (vi) $6 \cdot 2 \equiv 11 \cdot 2$ (mod 10), but $6 \not\equiv 11$ (mod 10).

✳

Comments on the Seminar Exercise.
  (i) T. We have that $117 - 0 = 9 \cdot 13$ so 117 is divisible by 9.
 (ii) T. By the division algorithm, $26 = 5 \cdot 5 + 1$, so $26 \equiv 1$ (mod 5).
(iii) F. In this example, the difference $-8 - (-3) = -5$, is not divisible by 11, so (iii) is false. Note that it is true that $-8 \equiv 3$ (mod 11) because $-8 - 3 = -11$ is divisible by 11.
 (iv) T. For, $-49 - (-7) = -49 + 7 = -42 = 6 \cdot (-7)$ is a multiple of 6.
  (v) T. We have $5 \cdot 3 = 15$, and $15 - 1 = 14 = 2 \cdot 7$.
 (vi) T. It is true that $6 \cdot 2 \equiv 11 \cdot 2$ (mod 10) because $12 - 22 = -10$ is divisible by 10. It is also true that $6 \not\equiv 11$ (mod 10) because $6 - 11 = -5$ is not

divisible by 10, Consequently, we *cannot* cancel the 2 in the congruence $6 \cdot 2 \equiv 11 \cdot 2$ (mod 10) and infer that $6 \equiv 11$ (mod 10).

※

Seminar/Classroom Activity. Let us take a moment to consider some practical congruence problems.

(i) If it is 10 o'clock now, what time will it be 35 hours from now? 100 hours from now? (Count mod 12.)

(ii) If today is Tuesday, what day of the week will it be 39 days from now? To solve this problem using congruences, what number do you want to take as modulus?

(iii) The switch for a lamp with a three way bulb has four positions: OFF, LOW LIGHT, MEDIUM LIGHT and HIGH LIGHT. Use the numbers 0, 1, 2 and 3 to represent these four positions. Suppose the switch is at 0 (the light is OFF), and is then turned through 6 positions. What type of light (LOW, MEDIUM or HIGH), if any, is on? Answer the same questions if the switch is at 0 and is turned through 13 positions. What is the modulus for this congruence problem?

(iv) What are some additional examples from everyday life where repetition of an action a certain number of times arrives back at the starting point. What are the moduli (the plural of "modulus" is "moduli") for these actions?

※

Comments on the Seminar/Classroom Activity.

(i) We add the integers 10 and 35 to obtain the sum $45 = 3 \cdot 12 + 9$. The hour hand starts at 0, winds all the way around passing 0 three times and then stops at 9. For 100 hours from now, we add 10 and 100 to obtain 110. The hour hand starts at 0, winds all the way around passing 0 nine times and then stops at 2.

(ii) Take 7 as modulus. The answer depends upon how you assign the numbers 0 through 6 to the days of the week. If you assign the numbers 0 through 6 to the days of the week, Sunday through Saturday, in the usual order, then Tuesday corresponds to 2 and $2 + 39 = 41 = 7 \cdot 5 + 6$. We have $41 \equiv 6$ (mod 7), and 39 days from now will be Saturday.

(iii) Take 4 as modulus. Then $0 + 6 = 6 = 1 \cdot 4 + 2$, so $6 \equiv 2$ (mod 4). Consequently, the light is MEDIUM. If the switch is turned through 13 positions, then $0 + 13 = 3 \cdot 4 + 1$, and the light is LOW.

※

Congruences with respect to a fixed modulus m have many of the same properties as equations. This might be expected since the congruence $a \equiv b$ (mod m) translates into an equation $a - b = km$ for some integer k.

But there are some surprises. Recall that if we have an *equation* of the form

$$a \cdot c = b \cdot c,$$

where a, b and c are integers and $c \neq 0$, then multiplicative cancellation holds. In other words, we may divide both sides of the equation by c to obtain the equality: $a = b$. However, if we have the *congruence*

$$a \cdot c \equiv b \cdot c \pmod{m},$$

where a, b and c are integers and $c \not\equiv 0 \pmod{m}$, it is not necessarily true that $a \equiv b \pmod{m}$, as we saw in the example in part (vi) of the Seminar Exercise.

Another surprise is that the product of two integers, each of which is not congruent to 0 mod m, can be congruent to 0 mod m. If we consider congruence mod 6, we have $2 \not\equiv 0 \pmod 6$ and $3 \not\equiv 0 \pmod 6$, but $2 \cdot 3 \equiv 0 \pmod 6$. Find another such example.

4.1. Properties of Congruence

Let m be an integer greater than 1. Let a, b, c and d be any integers. Note that the modulus m is *fixed* for each property stated below.

  (i) **Reflexive Property.** $a \equiv a \pmod{m}$.
 (ii) **Symmetric Property.** If $a \equiv b \pmod{m}$, then $b \equiv a \pmod{m}$.
(iii) **Transitive Property.** If $a \equiv b \pmod{m}$ and $b \equiv c \pmod{m}$, then $a \equiv c \pmod{m}$.
 (iv) **Additive Property.** If $a \equiv b \pmod{m}$ and $c \equiv d \pmod{m}$, then $(a + c) \equiv (b + d) \pmod{m}$.
  (v) **Multiplicative Property.** If $a \equiv b \pmod{m}$ and $c \equiv d \pmod{m}$, then $ac \equiv bd \pmod{m}$.
 (vi) **Powers Property.** If $a \equiv b \pmod{m}$, then $a^j \equiv b^j \pmod{m}$.

✳

Seminar Exercise. Try to verify each of the properties above before you read further. Remember that congruence is defined using actual equations. Nearly every basic property of congruence is proved by translating the statement into equations.

✳

To verify the **Reflexive Property**, we note that $a - a = 0$ and that 0 is divisible by m, so $a \equiv a \pmod{m}$ by the definition of congruence.

For the **Symmetric Property**, note that the hypothesis $a \equiv b \pmod{m}$ means that $a - b$ is divisible by m. So, we have $a - b = km$, for some integer k. It follows that $b - a = (-k)m$, so $b \equiv a \pmod{m}$.

Next consider the **Transitive Property**. The congruence $a \equiv b \pmod{m}$ means that $a - b = hm$, for some integer h. The congruence $b \equiv c \pmod{m}$

means that $b - c = km$, for some integer k. To show that $a \equiv c \pmod{m}$, we need an equation of the form $a - c = tm$, for some integer t. If we add the two equations given by the hypotheses: $a - b = hm$, and $b - c = km$, we have

$$a - b + b - c = hm + km.$$

Since $-b + b = 0$, we may write

$$a - b + b - c = a + 0 - c = a - c = hm + km.$$

By the distributive rule, $hm + km = (h + k)m$. Thus,

$$a - c = (h + k)m,$$

and we have $a - c = tm$, where $t = h + k$ is an integer. It follows that $a \equiv c \pmod{m}$.

The **Additive Property** follows in a similar way. We translate the hypotheses into equations. We have $a \equiv b \pmod{m}$ and $c \equiv d \pmod{m}$, so there are integers h and k such that $a - b = hm$ and $c - d = km$. Just as above, we add the equations to obtain $a - b + c - d = hm + km$. We use the rules of arithmetic for the integers to rearrange the lefthand side,

$$a - b + c - d = a + c - (b + d) = hm + km,$$

and finally, we use distributivity on the righthand side,

$$a + c - (b + d) = (h + k)m = tm,$$

where t is the integer $h + k$. This equation tell us that $a + c \equiv b + d \pmod{m}$.

Note. If $c = d$, then the additive property states that $a \equiv b \pmod{m}$ implies that $(a + c) \equiv (b + c) \pmod{m}$. This is a useful special case.

The hypotheses of the **Multiplicative Property** imply that $a - b = hm$, and $c - d = km$, for some integers h and k. We seek an equation showing that $ac - bd$ is a multiple of m. If we multiply the equation $a - b = hm$ by c and the equation $c - d = km$ by b we have,

$$ac - bc = hcm \qquad \text{and} \qquad bc - bd = kbm.$$

Consequently, when we add the two equations above, we have

$$ac - bc + (bc - bd) = hcm + kbm.$$

Next, using the rules of arithmetic, we simplify this equation,

$$ac - bc + bc - bd = (hc + kb)m$$
$$ac - bd = (hc + kb)m$$

Thus, $ac \equiv bd \pmod{m}$.

Note. If $c = d$, then the multiplicative property states that $a \equiv b \pmod{m}$ implies that $(ac) \equiv (bc) \pmod{m}$. This is a useful special case.

The **Powers Property** follows from the multiplicative property. To see this, we apply the multiplicative property repeatedly, each time taking $c = a$ and $d = b$. To start, we apply the multiplicative property to the congruence $a \equiv b \pmod{m}$, taking $c = a$ and $d = b$. The result is the congruence $a^2 \equiv b^2 \pmod{m}$. Next we apply the multiplicative property to the congruence $a^2 \equiv b^2 \pmod{m}$, taking $c = a$ and $d = b$. The result is the congruence $a^3 \equiv b^3 \pmod{m}$. In general, if we apply the the multiplicative property to the congruence $a^{j-1} \equiv b^{j-1} \pmod{m}$, taking $c = a$ and $d = b$, we obtain the congruence $a^j \equiv b^j \pmod{m}$.

Note that in the Powers Property, the modulus m remains fixed. In the property we discuss next, the Powers Congruent to 0 Property, the hypothesis $a \equiv 0 \pmod{m}$ produces a conclusion mod *powers of m*.

Powers Congruent to 0 Property. If $a \equiv 0 \pmod{m}$, then $a^j \equiv 0 \pmod{m^j}$, for every $j \geq 0$.

✳

Seminar Exercise. Verify the Powers Congruent to 0 Property.

✳

The exercise is straightforward and will be applied frequently in our discussion of the divisibility tests.

Comments on the Seminar Exercise. We have $a \equiv 0 \pmod{m}$. So, m divides a. Consequently, m^j divides a^j for all $j \geq 0$, or, in other words, $a^j \equiv 0 \pmod{m^j}$, for all such j.

✳

Seminar Exercise.

(i) Show that the following subtraction property holds.
    Subtraction Property. If $a \equiv b \pmod{m}$ and $c \equiv d \pmod{m}$, then $a - c \equiv b - d \pmod{m}$.
(ii) Show that the following substitution property holds.
    Substitution Property. If $ab \equiv c \pmod{m}$ and $b \equiv d \pmod{m}$, then $ad \equiv c \pmod{m}$.

✳

Comments on the Seminar Exercise. To verify the subtraction property, we translate the hypotheses into equations. We have $a \equiv b \pmod{m}$ and $c \equiv d \pmod{m}$ imply that $a - b = hm$ and $c - d = km$, for some integers h and k. To show that $a - c \equiv b - d \pmod{m}$, we seek an equation of the form $(a - c) - (b - d) = tm$, for some integer t. By subtracting the two equations derived from the hypotheses and applying the rules of arithmetic, we obtain

$$(a - b) - (c - d) = hm - km$$
$$(a - c) - (b - d) = (h - k)m,$$
$$(a - c) - (b - d) = tm$$

where $t = h - k$ is an integer.

To verify the substitution property, again we translate the hypotheses into equations. We have $ab \equiv c \pmod{m}$ and $b \equiv d \pmod{m}$ imply that $ab - c = hm$ and $b - d = km$, for some integers h and k. To show that $ad \equiv c$ $\pmod{m}$, we seek an equation of the form $ad - c = tm$, for some integer t. Using the rules of arithmetic and the equation $b = d + km$, we have

$$ab - c = hm$$
$$a(d + km) - c = hm$$
$$ad + akm - c = hm$$

We subtract akm from both sides of the equation above, and write

$$ad - c = hm + (-akm)$$
$$ad - c = (h - ak)m = tm,$$

where t is the integer $h - ak$.

✳

We have seen that multiplicative cancellation may not hold for congruences. The next observation gives an important instance where cancellation is valid.

Recall that two integers c and m are *relatively prime* if the greatest common divisor of c and m, $\gcd(c, m)$, is 1.

OBSERVATION 4.1. *Let m be an integer greater than 1. Let a, b and c be integers. Assume that c and m are relatively prime. If $ac \equiv bc \pmod{m}$, then $a \equiv b \pmod{m}$.*

The observation states, in other words, that we may cancel a factor c **if c is relatively prime to the modulus m**. To verify this observation, we rely one of the most useful facts about relatively prime integers. Observation 2.2 in Seminar 4 states (in different notation) that if $\gcd(c, m) = 1$ and if m divides the product kc, for some integer k, then m divides k.

How does this apply to verify the observation above? Since $ac \equiv bc$ $\pmod{m}$, we have that m divides $ac - bc$. By the distributive rule, m divides $(a - b)c$. Since m and c are relatively prime, it follows from Seminar 4, Observation 2.2 that m divides $a - b$. Thus, by the definition of congruence mod m, $a \equiv b \pmod{m}$.

4.2. Multiplicative Inverses mod m

Recall that a mutiplicative inverse of an integer h is an integer k, such that $hk = 1$. In other words, a multiplicative inverse k of h is a solution for x of the linear equation

$$hx = 1.$$

We have seen that 1 and -1 are the only integers that have multiplicative inverses in the integers.

✳

Seminar Question. What do you think is meant by a multiplicative inverse of h mod m?

✳

Spend some time on this before reading the comments below.

Comments on the Seminar Question. A *multiplicative inverse of h mod m* is a solution, for x, of the congruence

$$hx \equiv 1 \quad (\text{mod } m).$$

In Seminar 7, we discuss solutions of congruences and, in particular, the question of which integers have inverses mod m.

✳

Seminar Exercise. By experimentation, find which of the integers 1, 2, 3 and 4 have multiplicative inverses mod 5, and identify the inverses. Do the same for the integers 1, 2 and 3 mod 4.

✳

Comments on the Seminar Exercise. Each of the integers 1, 2, 3 and 4 has a multiplicative inverse (mod 5). We have $1 \cdot 1 \equiv 1$ (mod 5), $2 \cdot 3 \equiv 1$ (mod 5), $3 \cdot 2 \equiv 1$ (mod 5), and, finally, $4 \cdot 4 \equiv 1$ (mod 5). Thus, 1 has multiplicative inverse 1 (mod 5), 2 has multiplicative inverse 3 (mod 5), 3 has multiplicative inverse 2 (mod 5), and 4 has multiplicative inverse 4 (mod 5). For the (mod 4) part of the exercise, you will find that 1 and 3 have multiplicative inverses (mod 4), but 2 does not. What do you suspect determines whether an integer has a multiplicative inverse mod m?

Note. This topic continues in Seminar 7, Section 5.

5. Congruence and Tests for Divisibility

Congruence is applied throughout mathematics. Our first application reveals that the popular classroom tests for divisibility by numbers such as 3 and 9 have their basis in modular arithmetic.

Seminar 11 contains a detailed discussion of base 10, or decimal, arithmetic for fractions. For now, we use the fact, familiar to all of you, that a positive integer such as 352 has a base 10 expansion

$$352 = 3 \cdot 10^2 + 5 \cdot 10 + 2,$$

and, more generally, each positive integer n has a base 10 expansion

$$n = a_k 10^k + a_{k-1} 10^{k-1} + \cdots + a_j 10^j +$$
$$a_{j-1}(10)^{j-1} + \cdots + a_2 10^2 + a_1 10 + a_0,$$

where k is a nonnegative integer, and the integers a_i satisfy $0 \leq a_i \leq 9$. We will use this notation for the base 10 expansion of an integer n throughout this section. Many of you know the criteria for divisibility by 3 and 9, but you may not know why they they work. Modular arithmetic is perfectly suited to explain this. The first tests we consider are based on the fact that 2 and 5 are the prime divisors of 10, and its consequence that $10 \equiv 0$ (mod 2) and $10 \equiv 0$ (mod 5).

5.1. Divisibility by 2, 4 and 5

We begin with an example, the integer 1932. We expand it in base 10:

$$1932 = 1 \cdot 10^3 + 9 \cdot 10^2 + 3 \cdot 10 + 2.$$

We know that 1932 is an even number and clearly divisible by 2, but 1932 provides a fine first example of using modular arithmetic to learn why this is so. First, it follows from $2 \mid 10$ that $10 \equiv 0$ (mod 2). If we apply the arithmetic of congruences mod 2 to the base 10 expansion above, we have by the substitution rule and the other arithmetic rules for congruences that

$$1932 = 1 \cdot 10^3 + 9 \cdot 10^2 + 3 \cdot 10 + 2$$
$$1932 \equiv 1(0)^3 + 9(0)^2 + 3(0)^1 + 0 \quad (\text{mod } 2)$$
$$1932 \equiv 0 \quad (\text{mod } 2).$$

This tells us that 2 divides 1932, and that 1932 is an even number. As we know, any number with last digit divisible by 2 is itself divisible by 2 and now we know the reason why. Note that if we apply the same procedure as above to the number 1929, we have

$$1929 = 1 \cdot 10^3 + 9 \cdot 10^2 + 2 \cdot 10 + 9$$
$$1929 \equiv 1(0)^3 + 9(0)^2 + 2(0)^1 + 9 \quad (\text{mod } 2)$$
$$1929 \equiv 9 \quad (\text{mod } 2)$$
$$1929 \equiv 1 \quad (\text{mod } 2),$$

where the last congruence follows from the fact that $9 = 2 \cdot 4 + 1$. Thus, 1929 is not divisible by 2 because it is not congruent to 0 (mod 2).

Next we consider whether 1932 is divisible by $2^2 = 4$. For this, we will work mod 4. Since 2 divides 10, i.e., $10 \equiv 0$ (mod 2), it follows from the powers Congruent to 0 property that $10^2 \equiv 0$ (mod 2^2). Thus $10^2 \equiv 0$

(mod 4). But the fact that 4 divides 10^2 means that 4 divides 10^j, for all $j \geq 2$. Thus, we may apply the same procedure as above to reduce the base 10 expansion of 1932 (mod 4). We begin with the equation

$$1932 = 1(10)^3 + 9(10)^2 + 3(10) + 2,$$

and use the rules of modular arithmetic, to reduce it mod 4,

$$1932 \equiv 1(0) + 9(0) + (3(10) + 2) \pmod 4$$
$$1932 \equiv 0 + 0 + 32 \pmod 4.$$

We have $32 \equiv 0 \pmod 4$, since 4 divides 32. Consequently, $1932 \equiv 0$ (mod 4) and 4 divides 1932. This argument shows that an integer n is divisible by 4 exactly when the integer composed of the last two digits of n is divisible by 4. (It follows from the statement in the Seminar/Classroom Activity below that $2^3 = 8$ does not divide 1932, since $2^3 = 8$ does not divide 932.)

✳

Seminar/Classroom Activity. It is true that an integer n is divisible by $2^3 = 8$ precisely when the integer composed of the last three digits of n are divisible by 8. More generally, it is true that a number n with, say, $k \geq h$ digits is divisible by 2^h, precisely when the integer composed of the last h digits of n is divisible by 2^h. You and your class might enjoy exploring this idea.

✳

Since 5 divides 10, we have $10 \equiv 0 \pmod 5$. Methods similar to those used above give us a test for divisibility by 5. By the powers Congruent to 0 property, we deduce that $10^2 \equiv 0 \pmod{5^2}$, and similar statements hold for higher powers of 5.

Arguments analogous to those for 2 show that 5 divides a number n precisely when 5 divides the last digit of the number n, and $5^2 = 25$ divides n precisely when 25 divides the integer composed of the last two digits of n. More generally, as you might expect, it is true that a number n with, say, $k \geq h$ digits is divisible by 5^h, precisely when the integer composed of the last h digits of n is divisible by 5^h.

Thus, $n = 2975$, is divisible by 5 and by $5^2 = 25$, but not by $5^3 = 125$. For, we have $5 \mid 5$, and $25 \mid 75$, but $125 \nmid 975$.

✳

Seminar/Classroom Activity. It is true that an integer n is divisible by $5^3 = 125$ precisely when the integer composed of the last three digits of n are divisible by 125. More generally, it is true that a number n with, say, $k \geq h$ digits is divisible by 5^h, precisely when the integer composed of the last h digits of n is divisible by 5^h. This is another excellent investigation that you and your class might enjoy.

✳

Summary. We summarize the divisibility tests for powers of 2 and for powers of 5. Let n be an integer with $k \geq h$ digits.
(i) The integer n is divisible by 2^h, precisely when the integer composed of the last h digits of n is divisible by 2^h.

(ii) The integer n is divisible by 5^h, precisely when the integer composed of the last h digits of n is divisible by 5^h.

Remark. The divibility tests for 2 and for 5 are comparable. The reason is that 2 and 5 are the prime divisors of 10.

5.2. Divisibility by 3 and by 9

Next, we explore how the well known tests for divisibility by 3 and by 9 are derived. You will see that the familiar addition of digits occurs because $10 \not\equiv 0 \pmod 3$ and $10 \not\equiv 0 \pmod 9$, but $10 \equiv 1 \pmod 3$ and $10 \equiv 1 \pmod 9$.

Consider the number $258 = 2 \cdot 10^2 + 5 \cdot 10 + 8$. Since $10 \equiv 1 \pmod 3$, it follows by the multiplicative property of congruence (note that the modulus remains the same for this property), that $10^2 \equiv 1 \pmod 3$ and $10^3 \equiv 1 \pmod 3$. When we use the additive and multiplicative properties of congruences to reduce

$$258 = 2 \cdot 10^2 + 5 \cdot 10 + 8$$

mod 3, we have

$$258 \equiv 2(1) + 5(1) + 8 \pmod 3$$

By additivity,

$$258 \equiv 15 \pmod 3$$

and, since 15 is divisible by 3, it follows that

$$258 \equiv 0 \pmod 3.$$

Thus, 3 divides 258.

The same argument produces the analogous result for any positive integer n. As we observed, the fact that 3 and 9 divide $10 - 1 = 9$ means that $10 \equiv 1 \pmod 3$ and $10 \equiv 1 \pmod 9$. Moreover, the multiplicative property of congruence implies that $10^j \equiv 1 \pmod 3$ and $10^j \equiv 1 \pmod 9$. Thus, as before, we have a general test for divisibility by 3 and by 9.

OBSERVATION 5.1. *If*

$$n = a_k 10^k + a_{k-1} 10^{k-1} + a_{k-2} 10^{k-2} + \cdots + a_2 10^2 + a_1 10 + a_0,$$

then n is divisible by 3 if and only if the sum of its digits is divisible by 3.

For when we apply the arithmetic properties of congruences to reduce the equation above mod 3, we have

$$n \equiv a_k(1)^k + a_{k-1}(1)^{k-1} + a_{k-2}(1)^{k-2} + \cdots + a_2(1)^2 + a_1(1) + a_0 \pmod 3$$

and so
$$n \equiv a_k + a_{k-1} + a_{k-2} + \cdots + a_2 + a_1 + a_0 \quad (\text{mod } 3)$$
Thus, $n \equiv 0 \pmod 3$ precisely when $a_k + a_{k-1} + a_{k-2} + \cdots + a_2 + a_1 + a_0 \equiv 0$ (mod 3), i.e., n is divisible by 3 precisely when the sum of the digits of n is divisible by 3.

This exact statement holds when 3 is replaced by 9. Of course, an integer divisible by 9 is divisible by 3.

✻

Seminar Exercise.

(i) Test whether the following numbers are divisible by 3.

1056, 25986533, 3939393939393

(ii) Test whether the following numbers are divisible by 9.

775, 65801, 4365792

✻

Comment on the Seminar Exercise. We hope you did not find the sum of the digits to test whether or not 3939393939393 is divisible by 3. Some questions are posed to test your ability for think for yourself.

✻

Now that we understand that the tests for divisibility by 3 or by 9 derive from modular arithmetic, we eliminate the symbols and summarize the tests very simply.
Summary. We summarize the divisibility tests for 3 and 9. A positive integer n is divisible by 3 precisely when the sum of its digits is divisible by 3. A positive integer n is divisible by 9 precisely when the sum of its digits is divisible by 9.

5.3. Divisibility by 11

The basis of the divisibility test for 11 is the congruence $10 \equiv -1 \pmod{11}$. By the multiplicative property, it follows that
$$10^k \equiv (-1)^k \quad (\text{mod } 11).$$
Moreover, we know that $(-1)^2 = 1$, $(-1)^3 = -1$, and, more generally, -1 raised to an even power is 1 and -1 raised to an odd power is -1.

We put these facts to work as we consider the integer $1452 = 1(10)^3 + 4(10)^2 + 5(10) + 2$. When we reduce 1452 mod 11, using the arithmetic properties of congruence, we have
$$1452 \equiv 1(-1)^3 + 4(-1)^2 + 5(-1) + 2 \quad (\text{mod } 11).$$

It follows that

$$1452 \equiv 1(-1) + 4(1) + 5(-1) + 2 \pmod{11}.$$

For the general case, let n be any positive integer with base 10 expansion

$$n = a_k 10^k + a_{k-1} 10^{k-1} + a_{k-2} 10^{k-2} + \cdots + a_j 10^j + \cdots + a_2 10^2 + a_1 10 + a_0.$$

When we use the arithmetic properties of congruence to reduce n mod 11, we have

$$n \equiv a_k(-1)^k + a_{k-1}(-1)^{k-1} + \cdots + a_j(-1)^j +$$
$$a_{j-1}(-1)^{j-1} + \cdots + a_2(-1)^2 + a_1(-1) + a_0 \pmod{11}.$$

Thus, n is congruent to the *alternating sum* of its digits and, it follows that n is divisible by 11 precisely when the alternating sum of its digits is congruent to 0 mod 11. Note that the alternating sum displayed above begins with $+a_k$ if k is even and with $-a_k$ if k is odd.

We consider two more examples. First, the integer 378719 has six digits and base 10 expansion

$$378719 = 3 \cdot 10^5 + 7 \cdot 10^4 + 8 \cdot 10^3 + 7 \cdot 10^2 + 1 \cdot 10 + 9.$$

To test for divisibility by 11, we take the alternating sum of its digits:

$$-3 + 7 - 8 + 7 - 1 + 9 = 11.$$

Since the alternating sum of its digits is congruent to 0 (mod 11), i.e., the alternating sum is divisible by 11, our arguments show that 378719 is divisible by 11.

Next consider the integer 822160823. The alternating sum of its digits is

$$8 - 2 + 2 - 1 + 6 - 0 + 8 - 2 + 3 = 22$$

and is divisible by 11. Thus, 822160823 is divisible by 11.

Summary. We summarize the divisibility test for 11. A positive integer n is divisible by 11 precisely when the integer that is the alternating sum of the digits of n is divisible by 11.

Seminar 7

More Modular Arithmetic

1. Congruence Classes

Let m be an integer greater than 1. This section focuses on the construction of a number system with precisely m elements. In fact, we will see that the elements of this number system are classes of integers defined by congruence mod m. This idea is one of the many intriguing results of modular arithmetic. Here is how it is done.

We take, as modulus, a positive integer m greater than 1. If k is an integer, there are many integers congruent to k mod m. For example, if $m = 5$ and $k = 1$, the integers -4, 1, 6 and 11, are all congruent to 1 mod 5, because each of these integers has remainder 1 when divided by 5. We collect together all of the integers with remainder 1 when divided by 5, and form the class, denoted by $\overline{1}$, of all integers congruent to 1 mod 5 :

$$\overline{1} = \{\ldots, -14, -9, -4, 1, 6, 11, 16, \ldots\}.$$

Using the division algorithm, the possible remainders mod 5 are 0, 1, 2, 3 and 4. To each of these remainders, there is associated a class $\overline{0}$, $\overline{1}$, $\overline{2}$, $\overline{3}$ and $\overline{4}$.

The class $\overline{0}$ consists of all integers that have remainder 0 when divided by 5, i.e., $\overline{0}$ is the class of all multiples of 5.

The class $\overline{1}$ consists of all integers that have remainder 1 when divided by 5.

The class $\overline{2}$ consists of all integers that have remainder 2 when divided by 5.

The class $\overline{3}$ consists of all integers that have remainder 3 when divided by 5.

The class $\overline{4}$ consists of all integers that have remainder 4 when divided by 5. These classes are called *congruence classes* (mod 5).

⁂

Seminar/Classroom Activity. Let $m = 5$. Following the example of $\overline{1}$ above, write down seven elements in each of the classes $\overline{0}$, $\overline{2}$, $\overline{3}$ and $\overline{4}$.

⁂

Comments on the Seminar/Classroom Activity. We repeat $\overline{1}$ to highlight the patterns in the numbers.

$$\overline{0} = \{\ldots, -10, -5, 0, 5, 10, 15, 20, \ldots\}$$
$$\overline{1} = \{\ldots, -24, -19, -14, -9, -4, 1, 6, \ldots\}$$
$$\overline{2} = \{\ldots, -3, 2, 7, 12, 17, 22, 27, \ldots\}$$
$$\overline{3} = \{\ldots, -7, -2, 3, 8, 13, 18, 23, \ldots\}$$
$$\overline{4} = \{\ldots, -16, -11, -6, -1, 4, 9, 14, \ldots\}$$

Observe that, in any congruence class, every pair of elements differs by a multiple of 5.

✳

The same construction can be carried out for any modulus m, and any of the integers $0, 1, 2, \ldots, m-2, m-1$. Let r be one of these integers. All of the integers congruent to r (mod m) have the same remainder r when divided by m, so we collect them together into a class that we call a *congruence class* (mod m), and designate the class by $\overline{r}$.

For example, if $m = 2$, there are two congruence classes.

$$\overline{0} = \{\ldots, -4, -2, 0, 2, 4, 6, \ldots\}$$
$$\overline{1} = \{\ldots, -3, -1, 1, 3, 5, 7, \ldots\}$$

As you see, $\overline{0}$ is the class of even integers, i.e., the integers that have remainder 0 when divided by 2, and $\overline{1}$ is the class of odd integers, i.e., the integers that have remainder 1 when divided by 2.

If $m = 4$, there are four congruence classes.

$$\overline{0} = \{\ldots, -8, -4, 0, 4, 8, 12, \ldots\}$$
$$\overline{1} = \{\ldots, -7, -3, 1, 5, 9, 13, \ldots\}$$
$$\overline{2} = \{\ldots, -6, -2, 2, 6, 10, 14, \ldots\}$$
$$\overline{3} = \{\ldots, -5, -1, 3, 7, 11, 15, \ldots\}$$

$\overline{0}$ is the class of integers that have remainder 0 when divided by 4, $\overline{1}$ is the class of integers that have remainder 1 when divided by 4, $\overline{2}$ is the class of integers that have remainder 2 when divided by 4, and $\overline{3}$ is the class of integers that have remainder 3 when divided by 4.

The pattern is clear. There are m distinct congruence classes mod m: $\overline{0}, \overline{1}, \overline{2}, \overline{3}, \ldots, \overline{m-2}$, and $\overline{m-1}$, because there are m distinct nonnegative remainders when divided by m.

The integers $0, 1, 2, \ldots, m-2, m-1$ are called the *least residues* (mod m). The word *residue* or "what is left over" cues us that a least residue (mod m) is a remainder obtained using the division algorithm to divide by m. For example, if $m = 4$, the integers 0, 1, 2 and 3 are the four least residues mod 4.

If k is not a least residue, we define $\overline{k}$ as follows. $\overline{k}$ is the class of all integers that are congruent to k mod m. Thus, for example, if $m = 4$,

$$\overline{11} = \{\ldots, -13, -9 - 5, -1, 3, 7, 11, 15, \ldots\}.$$

Compare $\overline{11}$ with $\overline{3}$ above. It appears that these classes are the same. In fact, we shall see that $\overline{11} = \overline{3}$.

For each integer $m > 1$, we let $\mathbb{Z}_m$ be the collection of m congruence classes mod m :

$$\mathbb{Z}_m = \{\overline{0}, \ \overline{1}, \ \overline{2}, \ \ldots, \overline{m - 2}, \ \overline{m - 1}\}.$$

The letter $\mathbb{Z}$ reminds us that congruence classes are derived from the integers, $\mathbb{Z}$, and the m denotes the modulus.

✳

Seminar/Classroom Activity. Let $m = 8$.
(1) Name the 8 congruence classes and write down at least six nonnegative and five negative integers in each class. What are the least residues mod 8?
(2) Using (1), answer the following questions.
(i) Are there any integers that are not in one of the eight congruence classes mod 8? If so, name some of them.
(ii) Are there any integers in more than one congruence class? If so, give an example.
(iii) Take the integer 11 in $\overline{3}$. Show that the eleven integers you wrote down in $\overline{3}$ in Part (1) of the Activity are congruent to 11 (mod 8).

✳

Comments on the Seminar/Classroom Activity.
(1)

$$\overline{0} = \{\ldots, -40, -32, -24, -16, -8, 0, 8, 16, 24, 32, \ldots\}$$
$$\overline{1} = \{\ldots, -39, -31, -23, -15, -7, 1, 9, 17, 25, 33, \ldots\}$$
$$\overline{2} = \{\ldots, -38, -30, -22, -14, -6, 2, 10, 18, 26, 34, \ldots\}$$
$$\overline{3} = \{\ldots, -37, -29, -21, -13, -5, 3, 11, 19, 27, 35, \ldots\}$$
$$\overline{4} = \{\ldots, -36, -28, -20, -12, -4, 4, 12, 20, 28, 36, \ldots\}$$
$$\overline{5} = \{\ldots, -35, -27, -19, -11, -3, 5, 13, 21, 29, 37, \ldots\}$$
$$\overline{6} = \{\ldots, -34, -26, -18, -10, -2, 6, 14, 22, 30, 38, \ldots\}$$
$$\overline{7} = \{\ldots, -33, -25, -17, -9, -1, 7, 15, 23, 31, 39, \ldots\}$$

The least residues mod 8 are 0, 1, 2, 3, 4, 5, 6 and 7.
Our comments on part (2) of the Seminar/Classroom Activity are contained in the next observation. For a fixed modulus m, the collection $\mathbb{Z}_m$ of m

congruence classes (mod m) has three very important properties that we
have witnessed in our examples.

OBSERVATION 1.1. *Let m be an integer greater than 1. The collection
$\mathbb{Z}_m$ of m congruence classes (mod m) has the following three properties.*

(i) *Every integer is an element of one of the classes in the collection $\mathbb{Z}_m =
\{\overline{0}, \overline{1}, \ldots, \overline{m-1}\}$. In other words, every integer is congruent to one
of the integers $0, 1, \ldots, m-1$.*

(ii) *No integer is an element in more than one congruence class in $\mathbb{Z}_m$. In
other words, no integer is congruent to more than one of the integers
$0, 1, \ldots, m-1$.*

(iii) *If k is an integer, then*

$$\overline{k} = \overline{r},$$

where $k = mq + r$, with $0 \leq r < m$.

To verify (i), we show that every integer a is in some congruence class
mod m. Applying the division algorithm to divide a by m, we have

$$a = qm + r,$$

where $0 \leq r < m$. It follows that $a \equiv r \pmod{m}$ and a is an element of the
congruence class $\overline{r}$ which is one of the elements in the collection $\mathbb{Z}_m$.

For (ii), suppose that h and k are least residues mod m, and that there
is an integer a in $\overline{h}$ and in $\overline{k}$. We must show that $h = k$. By definition, $a \equiv h$
$\pmod{m}$ and $a \equiv k \pmod{m}$. By the Symmetric Property of congruence,
$h \equiv a \pmod{m}$. We have, therefore, $h \equiv a \pmod{m}$ and $a \equiv k \pmod{m}$,
so $h \equiv k \pmod{m}$ by the Transitive Property of congruence. Since h and k
are least residues, we must have $h = k$.

For (iii), we have $k = mq + r$, with $0 \leq r < m$, so $k \equiv r \pmod{m}$,
and, by definition of $\overline{r}$, k is an integer in $\overline{r}$. Moreover, if a is an integer in
$\overline{k}$, then $a \equiv k \pmod{m}$ and $k \equiv r \pmod{m}$ imply that $a \equiv r \pmod{m}$,
by the Transitive Property of congruence. Thus, every integer in the class
$\overline{k}$ is in the class $\overline{r}$. Suppose that b is an integer in the class $\overline{r}$. We have
$b \equiv r \pmod{m}$, and by the Symmetric Property, $r \equiv k \pmod{m}$. Thus, by
transitivity, $b \equiv k \pmod{m}$ and b is in $\overline{k}$. Consequently, every element in
the class $\overline{r}$, is in the class $\overline{k}$. It follows that $\overline{k} = \overline{r}$.

These properties show that for every modulus $m > 1$, the collection of
congruence classes $\mathbb{Z}_m$ partitions $\mathbb{Z}$ into nonintersecting classes defined by
congruence mod m.

$\overline{0}$	$\overline{1}$	$\overline{2}$	$\overline{3}$	$\cdots$ $\cdots$ $\cdots$	$\overline{m-2}$	$\overline{m-1}$

$$\longrightarrow \quad \mathbb{Z} \quad \longleftarrow$$

2. The Arithmetic of Congruence Classes

The collection $\mathbb{Z}_m$ is a number system because, as we shall see, it has two operations, namely addition and multiplication, that arise very naturally from the addition and multiplication defined on the integers.

2.1. Addition and Multiplication in $\mathbb{Z}_m$

We begin by defining addition and multiplication in $\mathbb{Z}_2$.

✳

Seminar/Classroom Activity. For this activity, we return to the example of the light switch. The collection $\mathbb{Z}_2$ has two elements, namely $\overline{0}$ and $\overline{1}$. Think of these elements as corresponding, respectively, to not flipping the light switch (or doing nothing) and flipping the switch. Think of addition as following one motion (or nonmotion) by another. Carry out the following activity. (In addition or multiplication tables, the sum or product, respectively, of a number in a row and a number in a column is placed at the intersection of the row and the column.)

(i) Fill in the addition table below for $\mathbb{Z}_2$. Use the example of the light switch. For example, flipping the light switch followed by flipping the light switch again is the same as no motion at all, so $\overline{1} + \overline{1} = \overline{0}$.

+	$\overline{0}$	$\overline{1}$
$\overline{0}$		
$\overline{1}$		

(ii) Tell whether or not $\mathbb{Z}_2$ has an additive identity. If it does, what is it?

✳

Comments on the Seminar/Classroom Activity. Here is the addition table for $\mathbb{Z}_2$.

+	$\overline{0}$	$\overline{1}$
$\overline{0}$	$\overline{0}$	$\overline{1}$
$\overline{1}$	$\overline{1}$	$\overline{0}$

We see from the table that $\overline{0}$ is the additive identity.

✳

We conclude from the table that *to add in $\mathbb{Z}_2$, first add in $\mathbb{Z}$ and then "bar it,"* i.e., take the class of the sum in $\mathbb{Z}_2$.

In mathematical language, if $\bar{a}$ and $\bar{b}$ are elements in $\mathbb{Z}_2$, then the rule for addition is

$$\bar{a} + \bar{b} = \overline{a + b}.$$

Multiplication on $\mathbb{Z}_2$ is also derived very naturally from multiplication of integers. (The example of the light switch is not useful for multiplication.) The definition of multiplication is analogous to that for addition. *To multiply in $\mathbb{Z}_2$, first multiply in $\mathbb{Z}$ and then "bar it,"* i.e., take the class of the product in $\mathbb{Z}_2$.

More precisely, if $\bar{a}$ and $\bar{b}$ are elements in $\mathbb{Z}_2$, then the rule for multiplication is

$$\bar{a} \cdot \bar{b} = \overline{a \cdot b}.$$

✳

Seminar/Classroom Activity.

(i) Fill in the multiplication table for $\mathbb{Z}_2$.

$\cdot$	$\bar{0}$	$\bar{1}$
$\bar{0}$		
$\bar{1}$		

(ii) Does $\mathbb{Z}_2$ have a multiplicative identity?

✳

Comments on the Seminar/Classroom Activity.

$\cdot$	$\bar{0}$	$\bar{1}$
$\bar{0}$	$\bar{0}$	$\bar{0}$
$\bar{1}$	$\bar{0}$	$\bar{1}$

It follows from the table that $\bar{1}$ is the multiplicative identity.

✳

There is nothing special about $\mathbb{Z}_2$. For any modulus m, we define addition and multiplication in the system $\mathbb{Z}_m$ of integers $\pmod m$ in exactly the same way. If $\bar{a}$ and $\bar{b}$ are elements of $\mathbb{Z}_m$, then addition in $\mathbb{Z}_m$ is defined by

$$\bar{a} + \bar{b} = \overline{a + b},$$

and multiplication in $\mathbb{Z}_m$ is defined by

$$\bar{a} \cdot \bar{b} = \overline{ab}.$$

In each case, add or multiply in $\mathbb{Z}$ first and then take the class of the sum or product, respectively, in $\mathbb{Z}_m$.

For example, if we take $\overline{2}$ and $\overline{3}$ in $\mathbb{Z}_5$, then, following the rule $\overline{a} + \overline{b} = \overline{a + b}$, we have,

$$\overline{2} + \overline{3} = \overline{2 + 3} = \overline{5} = \overline{0},$$

because $5 \equiv 0 \pmod{5}$. Note that $\overline{0}$ is the additive identity in $\mathbb{Z}_5$, and the equality $\overline{2} + \overline{3} = \overline{0}$ shows that $\overline{3}$ is the additive inverse for $\overline{2}$.

Similarly, with the same numbers, $\overline{2}$ and $\overline{3}$ in $\mathbb{Z}_5$, following the rule $\overline{a} \cdot \overline{b} = \overline{a \cdot b}$, we have

$$\overline{2} \cdot \overline{3} = \overline{2 \cdot 3} = \overline{6} = \overline{1},$$

because $6 \equiv 1 \pmod{5}$.

✳

Seminar/Classroom Activity.

  (i) Take $m = 4$. Write down addition and multiplication tables for $\mathbb{Z}_4$.
  (ii) Take $m = 10$. Write down addition and multiplication tables for $\mathbb{Z}_{10}$.

✳

Comments on the Seminar/Classroom Activity.
(i) Here is the addition table for $\mathbb{Z}_4$.

$+$	$\overline{0}$	$\overline{1}$	$\overline{2}$	$\overline{3}$
$\overline{0}$	$\overline{0}$	$\overline{1}$	$\overline{2}$	$\overline{3}$
$\overline{1}$	$\overline{1}$	$\overline{2}$	$\overline{3}$	$\overline{0}$
$\overline{2}$	$\overline{2}$	$\overline{3}$	$\overline{0}$	$\overline{1}$
$\overline{3}$	$\overline{3}$	$\overline{0}$	$\overline{1}$	$\overline{2}$

Note that $\overline{0}$ is the additive identity for $\mathbb{Z}_4$. Can you identify additive inverses for each of the four elements of $\mathbb{Z}_4$?

Here is the multiplication table for $\mathbb{Z}_4$.

$\cdot$	$\overline{0}$	$\overline{1}$	$\overline{2}$	$\overline{3}$
$\overline{0}$	$\overline{0}$	$\overline{0}$	$\overline{0}$	$\overline{0}$
$\overline{1}$	$\overline{0}$	$\overline{1}$	$\overline{2}$	$\overline{3}$
$\overline{2}$	$\overline{0}$	$\overline{2}$	$\overline{0}$	$\overline{2}$
$\overline{3}$	$\overline{0}$	$\overline{3}$	$\overline{2}$	$\overline{1}$

Observe that $\overline{2} \cdot \overline{2} = \overline{2 \cdot 2} = \overline{4} = \overline{0}$. Consequently, the square of the nonzero element $\overline{2}$ in $\mathbb{Z}_4$ is equal to $\overline{0}$. Frequently, the number systems $\mathbb{Z}_m$ exhibit very interesting behavior.

Part (ii) of this Activity is left to you. As you work on it, note that in mod 10 arithmetic, the sum and product of two integers mod 10 is the units digit of, respectively, their sum and product as integers. For example, $\overline{8} + \overline{9} = \overline{17} = \overline{7}$ mod 10, and $\overline{8} \cdot \overline{9} = \overline{72} = \overline{2}$ mod 10.

2.2. The Rules of Arithmetic of Congruence Classes

Let m be an integer greater than 1. As we have seen, for integers a and b, the definition of addition in $\mathbb{Z}_m$ is

$$\overline{a} + \overline{b} = \overline{a + b},$$

and the definition of multiplication in $\mathbb{Z}_m$ is

$$\overline{a} \cdot \overline{b} = \overline{a \cdot b}.$$

We must address the important question of whether the definitions of addition and multiplication in $\mathbb{Z}_m$ are consistent. What we mean by consistency is best illustrated with an example. We take the elements $\overline{2}$ and $\overline{3}$ in $\mathbb{Z}_6$. We have, by definition,

$$\overline{2} + \overline{3} = \overline{2 + 3} = \overline{5} \ \text{ in } \ \mathbb{Z}_6.$$

But 8 is one of the integers in $\overline{2}$ and 15 is an integer in $\overline{3}$, so, by Observation 1.1, (iii), $\overline{8} = \overline{2}$ and $\overline{15} = \overline{3}$. For the definition of addition of congruence classes to be consistent, we must have

$$\overline{8} + \overline{15} = \overline{5}.$$

By definition of addition $\overline{8} + \overline{15} = \overline{23}$. But, $23 = 3 \cdot 6 + 5$, so by Observation 1.1, (iii), $\overline{23} = \overline{5}$.

The problem in the general case is this. Suppose we want to add $\overline{a}$ and $\overline{b}$. The definition states that the sum is $\overline{a + b}$. However, we have seen that for any element h in $\overline{a}$, $\overline{h} = \overline{a}$, and for any element k in $\overline{b}$, $\overline{k} = \overline{b}$. Thus, for addition of congruence classes to be consistent or to be, as is often said, "well defined," we must show that

$$\overline{a} + \overline{b} = \overline{h} + \overline{k},$$

i.e., we must show that the sum will not change if we use different representatives for the congruence classes.

We have by definition,

$$\overline{a} + \overline{b} = \overline{a + b} \ \text{ and } \ \overline{h} + \overline{k} = \overline{h + k}.$$

So we must show that $\overline{a + b} = \overline{h + k}$. Since h is in $\overline{a}$, and k is in $\overline{b}$, we have $h \equiv a \pmod{m}$ and $k \equiv b \pmod{m}$. Consequently, by the Additive Property of congruences, it follows that $h + k \equiv a + b \pmod{m}$. Thus, by Observation 1.1, (iii), $\overline{h + k} = \overline{a + b}$.

Similarly, it can be shown that the definition of multiplication is consistent or well defined, that is if $\overline{a} = \overline{h}$, and $\overline{b} = \overline{k}$, then $\overline{ab} = \overline{hk}$. We will point out other places where we must be careful to maintain consistency.

Example. Let $m = 12$. Consider the elements $\overline{7}$ and $\overline{10}$ in $\mathbb{Z}_{12}$. We have $19 \equiv 7 \pmod{12}$ and $34 \equiv 10 \pmod{12}$. Thus,

$$\overline{7} + \overline{10} = \overline{17} \text{ and } \overline{19} + \overline{34} = \overline{53}.$$

It follows that $\overline{17} = \overline{53} = \overline{5}$, since $17 = 1 \cdot 12 + 5$ and $53 = 4 \cdot 12 + 5$. Similarly,

$$\overline{7} \cdot \overline{10} = \overline{70} \text{ and } \overline{19} \cdot \overline{34} = \overline{646}.$$

We have

$$\overline{70} = \overline{646} = \overline{10},$$

since $70 = 5 \cdot 12 + 10$ and $646 = 53 \cdot 12 + 10$. This example illustrates consistency of the definitions of addition and multiplication in $\mathbb{Z}_{12}$. It also illustrates that we can choose small representatives of each congruence class, in particular, the least residues, to make our computations more manageable.

Next, we examine the properties of addition and multiplication in $\mathbb{Z}_m$ that are analogous to the same properties in $\mathbb{Z}$.

Properties of Addition in $\mathbb{Z}_m$.

A1 Closure. If $\overline{a}$ and $\overline{b}$ are elements in $\mathbb{Z}_m$, then, by definition, $\overline{a} + \overline{b}$ is an element of $\mathbb{Z}_m$, that is, $\overline{a} + \overline{b}$ is the congruence class $\overline{a + b}$.

A2 Associative Property. If $\overline{a}$, $\overline{b}$ and $\overline{c}$ are elements of $\mathbb{Z}_m$, then

$$(\overline{a} + \overline{b}) + \overline{c} = \overline{a} + (\overline{b} + \overline{c}).$$

A3 Commutative Property. For elements $\overline{a}$ and $\overline{b}$ in $\mathbb{Z}_m$,

$$\overline{a} + \overline{b} = \overline{b} + \overline{a}.$$

A4 Additive Identity. There is a unique element $\overline{0}$ in $\mathbb{Z}_m$ such that

$$\overline{a} + \overline{0} = \overline{0} + \overline{a} = \overline{a},$$

for all elements $\overline{a}$ in $\mathbb{Z}_m$.

A5 Additive Inverse. For every element $\overline{a}$ in $\mathbb{Z}_m$, there is precisely one element, denoted $-\overline{a}$, such that

$$\overline{a} + (-\overline{a}) = -\overline{a} + \overline{a} = \overline{0}.$$

Example. We give an example of the additive inverse of an element of $\mathbb{Z}_m$, and show that consistency is maintained. Let $m = 6$, and let us find the additive inverse of $\overline{2}$ in $\mathbb{Z}_6$. Note that 4 is a least residue mod 6 such that $2 + 4 = 6$. It follows that $\overline{2} + \overline{4} = \overline{2 + 4} = \overline{6} = \overline{0}$. Consequently,

$$-\overline{2} = \overline{4}.$$

Observe that -2 is another integer, such that $\overline{2} + \overline{-2} = \overline{2 + (-2)} = \overline{0}$. Since $-2 - 4 = -6$ is divisible by 6, we have $-2 \equiv 4 \pmod 6$. Thus, -2 is an element of $\overline{4}$ and it follows from Observation 1.1 (iii) that

$$-\overline{2} = \overline{4} = \overline{-2}.$$

This example illustrates the consistency of the definition of the additive inverse of a congruence class. It also illustrates the fact that *the additive inverse of an element $\overline{a}$ in $\mathbb{Z}_m$ is the congruence class of $-a$*, that is,

$$-\overline{a} = \overline{(-a)}.$$

Properties of Multiplication in $\mathbb{Z}_m$.

M1 Closure. If $\overline{a}$ and $\overline{b}$ are elements in $\mathbb{Z}_m$, then, by definition, $\overline{a} \cdot \overline{b}$ is an element of $\mathbb{Z}_m$, that is, $\overline{a} \cdot \overline{b}$ is the congruence class $\overline{a \cdot b}$.

M2 Associative Property. If $\overline{a}$, $\overline{b}$ and $\overline{c}$ are elements of $\mathbb{Z}_m$, then

$$(\overline{a} \cdot \overline{b}) \cdot \overline{c} = \overline{a} \cdot (\overline{b} \cdot \overline{c}).$$

M3 Commutative Property. For elements $\overline{a}$ and $\overline{b}$ in $\mathbb{Z}_m$,

$$\overline{a} \cdot \overline{b} = \overline{b} \cdot \overline{a}.$$

M4 Multiplicative Identity. There is a unique element $\overline{1}$ in $\mathbb{Z}_m$ such that

$$\overline{a} \cdot \overline{1} = \overline{1} \cdot \overline{a} = \overline{a},$$

for all elements $\overline{a}$ in $\mathbb{Z}_m$.

D Distributive Property. If $\overline{a}$, $\overline{b}$ and $\overline{c}$ are elements of $\mathbb{Z}_m$, then

$$\overline{a} \cdot (\overline{b} + \overline{c}) = \overline{a} \cdot \overline{b} + \overline{a} \cdot \overline{c}.$$

Keep in mind that when we add and multiply in $\mathbb{Z}_m$, informally speaking, we add and multiply "under the bar" using the properties of arithmetic in $\mathbb{Z}$, and then apply the definitions of addition and multiplication in $\mathbb{Z}_m$. To verify the Distributive Property, first we use the definitions of addition and multiplication in $\mathbb{Z}_m$, to write:

$$\overline{a} \cdot (\overline{b} + \overline{c}) = \overline{a \cdot (b + c)}$$

Then, "under the bar,"' we apply the distributive property in $\mathbb{Z}$:

$$\overline{a \cdot (b + c)} = \overline{a \cdot b + a \cdot c}.$$

Next, we apply the definition of addition in $\mathbb{Z}_m$:

$$\overline{a \cdot b + a \cdot c} = \overline{a \cdot b} + \overline{a \cdot c}.$$

Finally, we apply the definition of multiplication in $\mathbb{Z}_m$:

$$\overline{a \cdot b} + \overline{a \cdot c} = \overline{a} \cdot \overline{b} + \overline{a} \cdot \overline{c}.$$

3. What is a Linear Congruence?

In this part of the seminar, we put modular arithmetic to work. We look at the objects in modular arithmetic that are analogous to linear equations in ordinary arithmetic. A linear congruence is an expression of the form

$$bx \equiv a \pmod{m},$$

where $m > 1$ is a modulus, a and b are integers, and x is a variable. A *solution of a linear congruence* is *an integer* x that satisfies the congruence. The expression $bx \equiv a \pmod{m}$ is similar to the familiar linear equation of the form $bx = a$ in ordinary arithmetic. (Linear equations of this form will be discussed in detail in Seminars 8 - 10, the seminars on fractions.) Linear equations of the form $bx = a$ have an integer solution only when b divides a. Whereas, as we shall see, linear congruences have solutions in many instances where b does not divide a.

Remark. We use the word "integer" as both a noun and an adjective, and call "an integer x that satisfies a linear equation" an *integer solution* to the linear equation. Elsewhere, you might see the word "integral" as the adjectival form of "integer."

Consider, for example, the linear equation $3x = 1$. It has no integer solutions. It has, as you know, a unique solution, $x = \frac{1}{3}$, that is a fraction. On the other hand, the linear congruence

$$3x \equiv 1 \pmod{5},$$

does have a solution. In fact, it has many solutions. The integers $x = 2, -3,$ 7 are three of them. How do we know this? We substitute these numbers for x, in turn, and check if the congruence is satisfied. If we set $x = 2$, we have $3 \cdot 2 = 6$ and, since $6 \equiv 1 \pmod{5}$, the integer $x = 2$ is a solution to the congruence $3x \equiv 1 \pmod{5}$. We leave it to you to check that -3 and 7 are also solutions. Remember, for congruences, the word "solution" means "integer solution."

4. Solving Linear Congruences

To solve a congruence

$$bx \equiv a \pmod{m},$$

where $m > 1$ is the modulus, a and b are integers, and x is a variable, means to find all (integer) solutions of the congruence. However, since we are working mod m, we consider solutions to be "different" only if they lie in distinct congruence classes $\pmod{m}$. We explain what we mean by this by considering the example

$$5x \equiv 4 \pmod{3}.$$

A solution to this congruence is an integer x that satisfies the congruence, i.e., an integer x such that $5x - 4$ is divisible by 3. Let us try to find a solution by experimentation. The integer 0 is not a solution because 3 does not divide -4, but the integer 2 is a solution because $5(2) - 4 = 10 - 4 = 6$ is divisible by 3.

✳

Seminar Exercise. Find two more solutions, one positive and one negative, to the congruence $5x \equiv 4 \pmod 3$. Is every integer congruent to 2 $\pmod 3$ a solution to $5x \equiv 4 \pmod 3$?

✳

Comments on the Seminar Exercise. In addition to the solution $x = 2$ that we found above, among the many other solutions are -1 and 5. Moreover, it is true that every integer congruent to 2 $\pmod 3$ is a solution to the congruence $5x \equiv 4 \pmod 3$. For, if $h \equiv 2 \pmod 3$, then $h = 2 + 3k$, for some integer k. The integer $2 + 3k$ is a solution because

$$5(2 + 3k) \equiv 10 + 15k \equiv 4 \pmod 3.$$

✳

Here is something interesting to observe. All solutions that we have found for the congruence $5x \equiv 4 \pmod 3$ are congruent to 2 $\pmod 3$. In fact, all solutions to $5x \equiv 4 \pmod 3$ are congruent to 2 $\pmod 3$, as we shall see in Observation 4.2.

First, we note that the congruence $5x \equiv 4 \pmod 3$ may be reduced, using the additive and multiplicative properties of congruences developed in Seminar 6, Section 4.1, as follows.

$$5 \equiv 2 \pmod 3; \ 4 \equiv 1 \pmod 3 \ \Rightarrow \ 5x - 4 \equiv 2x - 1 \pmod 3.$$

Consequently, an integer is a solution of the congruence $5x \equiv 4 \pmod 3$ precisely when that integer is a solution of the congruence $2x \equiv 1 \pmod 3$. This leads us to the following observation.

OBSERVATION 4.1. *Consider the congruence*

$$bx \equiv a \pmod m.$$

Suppose that a' is an integer with $a' \equiv a \pmod m$ and b' is an integer with $b' \equiv b \pmod m$. Then, an integer x is a solution to the congruence $bx \equiv a \pmod m$ precisely when x is a solution to the congruence $b'x \equiv a' \pmod m$.

Thus, for the example of the congruence $5x \equiv 4 \pmod 3$, we may replace 5 $\pmod 3$ by 2 $\pmod 3$, and 4 $\pmod 3$ by 1 $\pmod 3$ to reduce the congruence to $2x \equiv 1 \pmod 3$. The class of solutions of both congruences is the same.

We are particularly interested in congruences $bx \equiv a \pmod m$, where b and m are relatively prime. For, as the next observation shows, all solutions belong to the same congruence class. We regard such a congruence as having a unique solution.

OBSERVATION 4.2. *Suppose a, b and m are integers with m > 1.*
(i) The congruence

$$bx \equiv 1 \pmod{m}$$

has a solution precisely when b and m are relatively prime.
(ii) The congruence

$$bx \equiv a \pmod{m}$$

has a solution when b and m are relatively prime.
(iii) When b and m are relatively prime, all solutions to the congruence $bx \equiv a \pmod{m}$, *for* $a \geq 1$, *are congruent* $\pmod{m}$.

Observe that b and m are the important players in these problems, and a has a smaller role.

By Observation 4.1, we may assume that $b < m$. Let $d = \gcd(b, m)$. To verify (i), we assume, first, that b and m are relatively prime, i.e., that $d = 1$. Consequently, there are integers s and t such that

$$sb + tm = 1.$$

In this equation s is the coefficient of b, and t is the coefficient of m. The equation asserts that sb is congruent to 1 $(\text{mod } m)$ and that s is a solution to the congruence $bx \equiv 1 \pmod{m}$.

On the other hand, if s is a solution to the congruence $bx \equiv 1 \pmod{m}$, there is an integer t such that $bs = 1 + tm$. Thus, 1 is a linear combination, $1 = bs - tm$. Since d divides $bs - tm$, it follows that d divides 1, and $d > 0$ implies $d = 1$.

(ii) To show that the congruence $bx \equiv a \pmod{m}$ has a solution when b and m are relatively prime, we use part (i) of the observation to infer that the congruence $bx \equiv 1 \pmod{m}$ has a solution s. Thus $bs \equiv 1 \pmod{m}$, and it follows that $abs \equiv a \pmod{m}$. Consequently, as is a solution to the congruence $bx \equiv a \pmod{m}$.

(iii) If s and s' are two solutions to $bx \equiv a \pmod{m}$, then $m \mid bs - a$ and $m \mid bs' - a$. Consequently, $m \mid b(s - s')$. But b and m are relatively prime, so by Seminar 4, Observation 2.2, $m \mid (s - s')$, that is $s \equiv s' \pmod{m}$. Hence, all solutions are congruent to $s \pmod{m}$.

Example. We solve the congruence $5x \equiv 2 \pmod{7}$. (Observe that 5 and 2 are already reduced mod 7.) Since 5 and 7 are relatively prime, the congruence has a solution. There are integers s and t such that $5s + 7t = 1$. We apply the Coefficient Algorithm to calculate s and t as follows.

We rewrite the equation $5s + 7t = 1$ in the form

$$s \cdot 5 = 1 - t \cdot 7.$$

We look for a value of t between -3 and 3 such that 5 divides $1 - t \cdot 7$.

value of t	$1 - t \cdot 7$	5 divides $1 - t \cdot 7$?
1	$1 - 7 = -6$	No, $5 \nmid -6$.
-1	$1 + 7 = 8$	No, $5 \nmid 8$
2	$1 - 2 \cdot 7 = -13$	No, $5 \nmid -13$
-2	$1 + 2 \cdot 7 = 15$	Yes, $15 = 3 \cdot 5$.

Thus, $t = -2$ and $s = 3$, and

$$3 \cdot 5 - 2 \cdot 7 = 1.$$

The final step is to multiply through by 2 to obtain

$$6 \cdot 5 - 4 \cdot 7 = 2 \quad \text{or} \quad 5 \cdot 6 - 7 \cdot 4 = 2.$$

Thus, $5 \cdot 6 \equiv 2 \pmod 7$ and $x = 6$ is a solution to the congruence $5x \equiv 2 \pmod 7$. Moreover, since 5 and 7 are relatively prime, every solution is congruent to 6 $\pmod 7$.

✳

Seminar Exercise. Find a solution to each of the following congruences:
(i) $2x \equiv 3 \pmod{11}$,
(ii) $13x \equiv 1 \pmod{15}$.

✳

Comments on the Seminar Exercise.
(i) We have $\gcd(2, 11) = 1$, so the congruence $2x \equiv 3 \pmod{11}$ has a solution. There are integers s and t such that $2s + 11t = 1$. We could apply the coefficient algorithm to calculate s and t, but it is clear by inspection that a solution is $s = 6$ and $t = -1$. Consequently, $2 \cdot 6 + (-1) \cdot 11 = 1$ Thus, $2 \cdot 6 \equiv 1 \pmod{11}$, and after multiplying the congruence through by 3, we have $2 \cdot 18 \equiv 3 \pmod{11}$. It follows that $s = 18$ is a solution to the congruence $2x \equiv 3 \pmod{11}$. Since $\gcd(2, 11) = 1$, every solution of $2x \equiv 3 \pmod{11}$ is congruent to 18 $\pmod{11}$. But $18 \equiv 7 \pmod{11}$, so 7 is also a solution to $2x \equiv 3 \pmod{11}$. (Remember, it pays to take a few moments before beginning the solution process to inspect the congruence carefully. Had we done that for the congruence $2x \equiv 3 \pmod{11}$, we might have more discerned more quickly that $x = 7$ is a solution.

(ii) We have $\gcd(13, 15) = 1$, and we use the coefficient algorithm to find s and t such that

$$13s + 15t = 1.$$

We rewrite this equation in the form

$$13s = 1 - 15t.$$

We test values of t between -7 and 7.

value of t	$1 - 15t$	13 divides $1 - 15t$?
1	$1 - 15 = -14$	No, $13 \nmid -14$
-1	$1 - (-15) = 16$	No $13 \nmid 16$
2	$1 - 30 = -29$	No, $13 \nmid -29$.
-2	$1 + 30 = 31$	No, $13 \nmid 31$
3	$1 - 45 = -44$	No, $13 \nmid -44$
-3	$1 + 45 = 46$	No, $13 \nmid 46$
4	$1 - 60 = -59$	No, $13 \nmid -59$
-4	$1 + 60 = 61$	No, $13 \nmid 61$
5	$1 - 75 = -74$	No, $13 \nmid -74$
-5	$1 + 75 = 76$	No, $13 \nmid 76$
6	$1 - 90 = -89$	No, $13 \nmid -89$
-6	$1 + 90 = 91$	Yes, $91 = 13 \cdot 7$.

Thus, $t = -6$ and $s = 7$. Consequently, $s = 7$ is a solution to the congruence $13x \equiv 1 \pmod{15}$. Since 13 and 15 are relatively prime, all solutions of this congruence are congruent to 7 (mod 15).

BUT WAIT! Take a look at the Euclidean Algorithm for computing $\gcd(13, 15)$:

$$15 = 13 \cdot 1 + 2$$
$$13 = 2 \cdot 6 + 1$$
$$6 = 1 \cdot 6 + 0.$$

In this case, the Reverse Euclidean Algorithm is a quick way to find s and t. We solve for 1 in the second equation. Then we use the first equation to substitute $15 - 13 \cdot 1$ for 2 in the second equation:

$$1 = 13 - 2 \cdot 6 = 13 - (15 - 13 \cdot 1)6.$$

After we distribute and gather terms, we have

$$1 = 7 \cdot 13 - 6 \cdot 15.$$

Consequently, it follows that $s = 7$ and $t = -6$. This simple solution demonstrates that perspicacity pays.

✳

5. Inverses mod m

In Seminar 6, Section 4.2, the multiplicative inverse of an integer b (mod m) was defined to be a solution of the congruence

$$bx \equiv 1 \pmod{m}.$$

We know by Observation 4.2 that b *has a multiplicative inverse* (mod m) *precisely when b and m are relatively prime.* Thus, 6 is a multiplicative inverse of 2 (mod 11) and -5 is a multiplicative inverse of 4 (mod 21). In place of -5, we may take the least residue 16 as a multiplicative inverse of 4 (mod 21), if we wish.

✳

Seminar Exercise. In this exercise, we ask you to investigate positive integers $a < m$ that are relatively prime to m and are their own multiplicative inverses mod m, i.e., all integers a that satisfy $a \cdot a \equiv 1 \pmod{m}$.

(i) Take $m = 8$. Find all integers less than 8 that are relatively prime to 8 and are their own multiplicative inverses mod 8.

(ii) For any integer m, find two examples of positive integers less than m that are their own multiplicative inverses mod m.

✳

Comments on the Seminar Exercise.

(i) The positive integers less than 8 that are relatively prime to 8 are 1, 3, 5, and 7. If you square each of these numbers, you will find that each of these squares is congruent to 1 (mod 8).

(ii) This is left as a challenge for you. Try examples first.

Seminar 8

The Arithmetic of Fractions

1. Introduction

One of the most perplexing topics in the teaching of arithmetic to elementary and middle grade students is addition, multiplication and division of fractions. Students are confused, and their parents complain that there are too many rules and that their children do not know which rule to apply when.

We have dedicated five of the seminars in this guide to the study of fractions. We look at the whole range of concepts involved (addition, multiplication and division of fractions, common denominators, equivalent fractions, decimals, order, etc.) and suggest to you a slightly different arrangement of topics. For example, we propose teaching multiplication of fractions before addition. With the topic of multiplication in hand, common denominators and equivalent fractions are more readily understood and are available for use when discussing the knotty problem of addition.

We begin this seminar with a question you might ask your students.

2. Fractions and Why We Need Them

*

Seminar/Classroom Discussion. Why do we need fractions; why aren't the integers sufficient?

*

Comments on the Seminar/Classroom Discussion. Perhaps you had some of the following thoughts.

(i) Positive integers are ideal for counting, but are far from ideal for measuring. There are an integer number of students in your class, but it is highly unlikely that the height of each student can be measured precisely in integer inches. Thus, new numbers are required, numbers that are larger than a particular integer but smaller than that integer plus 1.

(ii) Another reason we need fractions is to describe parts or pieces of a whole. How else do we specify our share of a collection of objects?

(iii) The final idea we mention here is that of solving equations of the form $bx = a$, where a and b are integers with $b \neq 0$.

✳

We elaborate on point (iii). Suppose that a and b are integers with b nonzero. When does the equation

$$bx = a$$

have an integer solution? This equation rarely has a solution that is an integer. The solution is an integer only if b divides a, that is, only if there is an integer k such that $a = kb$. Consider, for example, the equations $3x = 15$ and $3x = 10$. Since $3 \cdot 5 = 15$, the integer $x = 5$ is a solution to the first equation. However, the equation $3x = 10$ does not have an integer solution because 3 does not divide 10. If we seek a number system containing solutions to *all* equations of the form $bx = a$ (with $b \neq 0$), not just when b divides a, we must enlarge the system of integers. A number in this new larger system is called a *fraction*. A fraction is written

$$\frac{a}{b},$$

where a and b are integers and $b \neq 0$. In the course of this seminar, you will see that fractions are the numbers required for more precise measuring, and for describing parts or pieces of a whole.

We say that two fractions $\frac{a}{b}$ and $\frac{c}{d}$ are *equal* precisely when $a = c$ and $b = d$. The integer a is called the *numerator* of the fraction and the integer b is called the *denominator* of the fraction. The fraction $\frac{a}{b}$ may also be written a/b. (We use the "slash" notation primarily for in line fractions and the "bar" notation for displayed fractions.)

Some examples illustrating that a fraction a/b is a solution of the equation $bx = a$, where a and b are integers, with $b \neq 0$, follow. The fraction $2/3$ is a solution of the equation $3x = 2$; the fraction $-1/9$ is a solution of the equation $9x = -1$; and $52/25$ is a solution of the equation $25x = 52$. (Note that we do not require that the numerator of a fraction be less than its denominator. This will be discussed in Seminar 9 when we take up mixed numbers.)

We consider the integers to be a subset of the set of fractions. If n is an integer, we identify n with the fraction $n/1$, where $n/1$ is a solution of the equation $1x = n$. Thus, 3 is identified with the fraction $3/1$, -5 is identified with the fraction $-5/1$ and 2231 is identified with $2231/1$. In particular, the integer 0 is identified with the fraction $0/1$, and the integer 1 is identified with the fraction $1/1$.

Note. Henceforth, the words "a/b is a fraction," mean that a and b are integers and that $b \neq 0$.

3. Multiplication of Fractions and Common Denominators

3.1. Multiplication of Fractions

We begin this section with the familiar rule for multiplication of fractions. You recall that if a/b and c/d are fractions, the rule for finding their product is

$$\frac{a}{b} \cdot \frac{c}{d} = \frac{a \cdot c}{b \cdot d} = \frac{ac}{bd}.$$

Note that ac and bd are integers and that b and d nonzero implies that bd is nonzero by Property (NZ), see Seminar 2, Section 5. Thus, it follows that the product ac/bd is a fraction. Observe that the multiplication rule for fractions is derived from the operation of multiplication of integers.

Let us look at three examples.

(i)
$$\frac{1}{4} \cdot \frac{3}{10} = \frac{1 \cdot 3}{4 \cdot 10} = \frac{3}{40}.$$

(ii)
$$\frac{3}{5} \cdot \frac{7}{4} = \frac{3 \cdot 7}{5 \cdot 4} = \frac{21}{20}.$$

(iii)
$$\frac{-2}{3} \cdot \frac{11}{8} = \frac{(-2) \cdot 11}{3 \cdot 8} = \frac{-22}{24}.$$

✳

Seminar/Classroom Discussion. How do you motivate multiplication of fractions for your students?

✳

Comments on the Seminar/Classroom Discussion. You might do what we often do, namely, discuss taking "a part of a part." Suppose we cut a "filled-in circle," or disc, out of colored paper, cut it in half and then cut one half in half. When students are asked what part of the whole circle is one half of one half, they instinctively know that each half of a half is a quarter of the circle. Written in mathematical language, we have

$$\frac{1}{2} \cdot \frac{1}{2} = \frac{1 \cdot 1}{2 \cdot 2} = \frac{1}{4}.$$

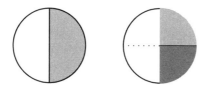

Next, suppose we cut the circle into thirds, and then cut one third in half. Frequently, when students are asked what part of the whole circle is one half of one third, some are unsure.

When we cut each of the other two thirds in half, and the students observe the six sections making a whole circle, they know that one half of one third is one sixth. Once again, when we write this down mathematically, we have a multiplication rule for calculating a part of a part:

$$\frac{1}{2} \cdot \frac{1}{3} = \frac{1 \cdot 1}{2 \cdot 3} = \frac{1}{6}.$$

✳

Problems of finding a part of a part can be solved by multiplication of fractions. Thus, a part of a part means

fraction · fraction.

This motivation for multiplication of fractions by taking a part of a part involves fractions with positive numerator and denominator and with numerator smaller than the denominator. However, the multiplication rule for fractions holds for all fractions. To mention a few of the many cases, the multiplication rule holds, for example, when the numerator is greater than the denominator or the numerator is negative:

$$\frac{56}{7} \cdot \frac{-12}{13} = \frac{56 \cdot (-12)}{7 \cdot 13} = \frac{-672}{91}.$$

The multiplication rule for fractions is based entirely on multiplication of integers. It follows directly from the properties of multiplication of integers that multiplication of fractions has the properties of closure, associativity and commutativity. The existence of a multiplicative identity will be discussed later. There is one property of multiplication of fractions, namely the existence of multiplicative inverses for nonzero fractions, that the integers do not have. We will also discuss this property in a subsequent section.

3.2. Common Denominators

Here is a very interesting question. Is it possible to convert any two fractions a/b and c/d to fractions with the same denominator using the operation of multiplication defined above? Before we do anything else, let us find out why this question is interesting. Suppose I have a plate of 15 pancakes, hot off the griddle. Suppose you are permitted to take 1/3 of the pancakes and I am allowed to have 2/5 of the pancakes. How many pancakes may each of us have? Consider the fractions 1/3 and 2/5. It is not immediately clear how to translate these fractions into numbers of pancakes. So let us put the fractions into a form in which they can easily be compared to one another. We use the fact that the product of the denominators is 15, the number of pancakes. Consequently, if we multiply the fraction $1/3$ by the fraction $5/5$, the product is the fraction $5/15$:

$$\frac{1}{3} \cdot \frac{5}{5} = \frac{1 \cdot 5}{3 \cdot 5} = \frac{5}{15},$$

and if we multiply the fraction $2/5$ by the fraction $3/3$, the product is $6/15$:

$$\frac{2}{5} \cdot \frac{3}{3} = \frac{2 \cdot 3}{5 \cdot 3} = \frac{6}{15}.$$

Both $1/3$ and $2/5$ have been converted by means of multiplication to fractions with a *common denominator*, namely 15. Although the fractions $1/3$ and $5/15$ are not equal, and the fractions $2/5$ and $6/15$ are not equal, both pairs of fractions are what we will call *equivalent fractions* when we define the term in Section 4.1. The fractions $5/15$ and $6/15$ specify exactly how many pancakes each of us may have, namely, 5 for you and 6 for me!

This method of finding a common denominator works for any two fractions a/b and c/d. We multiply a/b by the fraction with numerator and denominator equal to the denominator d of the fraction c/d :

$$\frac{a}{b} \cdot \frac{d}{d} = \frac{ad}{bd},$$

and we multiply c/d by the fraction with numerator and denominator equal to the denominator b of the fraction a/b :

$$\frac{c}{d} \cdot \frac{b}{b} = \frac{cb}{db} = \frac{bc}{bd}.$$

Thus, we have converted the fractions a/b and c/d to two fractions with common denominator equal to the product bd.

Suppose, for example, we require a common denominator for the fractions $5/11$ and $-7/10$. We multiply $5/11$ by $10/10$:

$$\frac{5}{11} \cdot \frac{10}{10} = \frac{50}{110}$$

and we multiply the fraction $-7/10$ by $11/11$:

$$\frac{-7}{10} \cdot \frac{11}{11} = \frac{-77}{110}.$$

Thus, both fractions have a common denominator equal to the product of their original denominators.

Next, we show you an enhanced version of the method above. Consider the fractions $5/6$ and $1/14$. We could apply the method above and convert $5/6$ and $1/14$ to fractions with common denominator $6 \cdot 14 = 84$. However, observe that $42 = 6 \cdot 7$ and $42 = 14 \cdot 3$. So 42 is also a multiple of the denominator 6 and of the denominator 14. In other words, 42 is a common multiple of the denominators. (In fact, if you examine the prime factorizations of 6 and 14, you will see that 42 is the least common multiple of 6 and 14.) In this example, we aim for two fractions with denominator 42. We multiply the fraction 5/6 by the fraction 7/7, and we multiply the fraction 1/14 by the fraction 3/3. (Observe that to transform two fractions to fractions with a common denominator, we always multiply by fractions of the form k/k.) We have

$$\frac{5}{6} \cdot \frac{7}{7} = \frac{35}{42} \quad \text{and} \quad \frac{1}{14} \cdot \frac{3}{3} = \frac{3}{42}.$$

Thus, we have a common denominator 42 that is less than the product 84 of the two denominators.

More generally, we show that if a/b and c/d are any fractions and if n is any nonzero integer that is a common multiple of b and d, then we can find a common denominator in a similar manner. Suppose that

$$n = hb \text{ and } n = kd,$$

for some nonzero integers h and k. We know that $hb = n$, so if we multiply a/b by h/h, the product fraction will have denominator n. Also, we know that $kd = n$, so if we multiply c/d by k/k, the product fraction will have denominator $kd = n$:

$$\frac{a}{b} \cdot \frac{h}{h} = \frac{ah}{bh} = \frac{ah}{n} \quad \text{and} \quad \frac{c}{d} \cdot \frac{k}{k} = \frac{ck}{dk} = \frac{ck}{n}.$$

Thus, we have the fractions ah/n and ck/n with common denominator n, which are "equivalent" to a/b and c/d, respectively.

There are many ways to convert two fractions a/b and c/d to fractions with a common denominator. For their relevance to the arithmetic of fractions, the two most important choices for a common denominator are:

(i) the product bd of the denominators;
(ii) the least common denominator, studied in the next section.

3.3. Two Algorithms for Finding the Least Common Denominator

In this section, we assume that the denominator f of every fraction e/f is positive. Of course, its numerator e may be any integer. This assumption will be justified in the next section when we discuss equivalence.

Among all common denominators of two fractions a/b and c/d, there is a smallest positive denominator. It is called the *least common denominator* or the *lowest common denominator* and is denoted by

$$\text{lcd}\left(\frac{a}{b}, \frac{c}{d}\right).$$

In fact, see Section 3 of Seminar 5, the least common denominator of a/b and c/d is equal to the least common multiple of b and d :

$$\text{lcd}\left(\frac{a}{b}, \frac{c}{d}\right) = \text{lcm}[b, d].$$

Each of the two methods for calculating lcm[b, d], discussed in Section 3 of Seminar 5, furnishes us with a method for calculating lcd($a/b, c/d$).

(i) LCD by Prime Factorization.

The first method depends on the prime factorizations of b and d. We discovered, in Seminar 5, that *lcm[b, d] is the product of the primes appearing in the factorization of b or the factorization of d, with the exponent of each prime in this product equal to the **maximum** number of times it appears in one factorization or the other.*

For example, let us look at $a/b = 25/72$ and $c/d = 31/84$. The prime factorization of 72 is $72 = 2^3 \cdot 3^2$, and the prime factorization of 84 is $84 = 2^2 \cdot 3 \cdot 7$. Thus, by the rule italicised above, lcm[$72, 84$] $= 2^3 \cdot 3^2 \cdot 7 = 504$. Consequently, lcd($25/72, 31/84$) $= 504$.

(ii) LCD by GCD.

The second method uses the formula

$$\text{lcd}\left(\frac{a}{b}, \frac{c}{d}\right) = \text{lcm}[b, d] = \frac{bd}{\gcd(b, d)}.$$

See Section 3 of Seminar 5. The $\gcd(b, d)$ may be calculated by means of the Euclidean Algorithm, see Seminar 3, or by means of prime factorization, see Seminar 5. We consider the same example as above: $a/b = 25/72$ and $c/d = 31/84$. We compute $\gcd(72, 84)$ by the Euclidean Algorithm:

$$84 = 1 \cdot 72 + 12$$
$$72 = 6 \cdot 12 + 0.$$

Thus, $\gcd(72, 84) = 12$. Next, we apply the formula above

$$\text{lcd}\left(\frac{25}{72}, \frac{31}{84}\right) = \frac{72 \cdot 84}{12} = 504.$$

To convert the fractions $25/72$ and $31/84$ to fractions with least common denominator 504, we write $504 = 72 \cdot 7$, and $504 = 84 \cdot 6$. Thus $25/72$ converts to

$$\frac{25}{72} \cdot \frac{7}{7} = \frac{175}{504}$$

and $31/84$ converts to

$$\frac{31}{84} \cdot \frac{6}{6} = \frac{186}{504}.$$

As we shall see in Seminar 10, it is now straightforward to add the fractions. Thus, we have two algorithms for computing the least common denominator, one by prime factorization, the other by the Euclidean Algorithm.

More Examples.

(i) Consider the fractions $5/12$ and $7/18$. By the Euclidean Algorithm or by prime factorization, $\gcd(12, 18) = 6$, and $\text{lcd}(5/12, 7/18) = (12 \cdot 18)/6 = 36$.

Thus, 5/12 converts to 15/36 and 7/18 converts to 14/36.

(ii) Let us calculate lcd(17/198, 5/252). Here, the denominators are rather large integers, so we find lcd(17/198, 5/252) by finding gcd(198, 252) using the Euclidean Algorithm. We have

$$252 = 1 \cdot 198 + 54$$
$$198 = 3 \cdot 54 + 36$$
$$54 = 1 \cdot 36 + 18$$
$$36 = 2 \cdot 18 + 0.$$

Consequently, gcd(198, 252) = 18, and

$$\text{lcd}\left(\frac{17}{198}, \frac{5}{252}\right) = \text{lcm}[198, 252] = \frac{198 \cdot 252}{18} = 2772.$$

To convert the given fractions to fractions with this denominator, we write $2772 = 198 \cdot 14$ and $2772 = 252 \cdot 11$. Thus, 17/198 converts to

$$\frac{17}{198} \cdot \frac{14}{14} = \frac{238}{2772}$$

and 5/252 converts to

$$\frac{5}{252} \cdot \frac{11}{11} = \frac{55}{2772}.$$

✲

Seminar Exercise. Find a common denominator for each pair of fractions listed below. Before you begin to write anything, think about which method you are going to use and why. Finally, write each pair of fractions as a pair with least common denominator.

  (i) 1/4 and 6/15
 (ii) 2/3 and 5/6
(iii) 101/102 and 25/222

✲

Comments on the Seminar Exercise.

  (i) We choose to compute the least common denominator for this example by prime factorization. We have $4 = 2^2$ and $15 = 3 \cdot 5$. The factorizations have no primes in common so lcm$[4, 15] = 4 \cdot 15$. Consequently,

$$\text{lcd}\left(\frac{1}{4}, \frac{6}{15}\right) = \text{lcm}[4, 15] = 4 \cdot 15 = 60.$$

We have $60 = 4 \cdot 15$, so

$$\frac{1}{4} \cdot \frac{15}{15} = \frac{15}{60},$$

and

$$\frac{6}{15} \cdot \frac{4}{4} = \frac{24}{60}.$$

(ii) For the fractions $2/3$ and $5/6$, observe that 6 is a multiple of 3, so $6 = \mathrm{lcm}[3, 6]$ is the least common multiple of the denominators of both fractions. We have $5/6$ with denominator 6 and

$$\frac{2}{3} \cdot \frac{2}{2} = \frac{4}{6}.$$

(iii) The denominators 102 and 222 are rather large, so we choose to compute $\mathrm{lcd}(101/102, 25/222)$ by the Euclidean Algorithm.

$$222 = 2 \cdot 102 + 18$$
$$102 = 5 \cdot 18 + 12$$
$$18 = 1 \cdot 12 + 6$$
$$12 = 2 \cdot 6 + 0$$

Thus $\gcd(102, 222) = 6$. To compute $\mathrm{lcm}[102, 222]$ we divide the product $102 \cdot 222 = 22644$ by 6, and we conclude that $\mathrm{lcm}[102, 222] = 3774$. Observe that $3774 = 102 \cdot 37$, and $11322 = 222 \cdot 17$. Thus,

$$\frac{101}{102} \cdot \frac{37}{37} = \frac{3737}{3774}.$$
$$\frac{25}{222} \cdot \frac{17}{17} = \frac{425}{3774}.$$

✳

4. Equivalence of Fractions

4.1. Definition of Equivalence

The fundamental idea discussed in the last section is that any two fractions can be multiplied by suitable fractions of the form k/k, where k is a nonzero integer, to produce two fractions with a common denominator. What we did not explain completely there is the relation of each fraction to the corresponding modified fraction. The question is: In what sense are these fractions "the same?" We explain it in detail in this section.

Recall that two fractions a/b and c/d are *equal* precisely when $a = c$ and $b = d$. The fractions $2/5$ and $4/10$ are not equal, but it is obvious that they are related.

For example, suppose I have 30 baseball cards that I offer to you. If I break up the cards into 5 groups of 6 cards each, then each group represents $1/5$ of the cards. If you select 2 of the groups, you have $2/5$ of the cards, that is, 12 cards. On the other hand, if I break up the cards into 10 groups of 3 cards each, and you select four of these groups, you have $4/10$ of the cards, again, 12 cards. Thus, the fractions $2/5$ and $4/10$ represent the same share of the collection of baseball cards, namely 12 of 30. We will call two fractions equivalent if they represent the same share of a collection of objects or the same part of a whole. The formal definition of equivalent fractions follows.

Two fractions a/b and c/d are said to be *equivalent* fractions if, there are nonzero integers h and k so that

$$\frac{a}{b} \cdot \frac{h}{h} = \frac{c}{d} \cdot \frac{k}{k}.$$

If we compute the products on both sides of the equation above and apply the definition of equality of fractions, we have that two fractions a/b and c/d are equivalent if there are nonzero integers h and k, such that

$$ah = ck \text{ and } bh = dk.$$

If a/b is equivalent to c/d, we write $a/b \sim c/d$.

We have seen that, for any fractions a/b and c/d, nonzero integers h and k always can be found so that $bh = dk$. *What is new here is that for the fractions to be equivalent, equality must also hold for the numerators, that is, the equality $ah = ck$ must hold.*

Examples.

(i) We check that $2/5$ and $4/10$ satisfy the definition of equivalence. For this pair of fractions, we take $h = 2$ and $k = 1$. Then,

$$\frac{2}{5} \cdot \frac{2}{2} = \frac{4}{10} \quad \text{and} \quad \frac{4}{10} \cdot \frac{1}{1} = \frac{4}{10}.$$

Thus, we have equality of the modified fractions, so $2/5 \sim 4/10$.

(ii) On the other hand, observe that the fractions $1/2$ and $2/3$ can be written with common denominator 6. The fraction $1/2$ converts to $3/6$ and the fraction $2/3$ converts to $4/6$. However, the fractions $3/6$ and $4/6$ are *not equal*, so $1/2$ and $2/3$ are *not equivalent*.

(iii) We identified the fraction $1/1$ with the integer 1. It is immediate, from the definition, that the fractions $1/1, -1/-1, 2/2, -2/-2, 3/3, -3/-3, \ldots$ are equivalent to $1/1$, as are all fractions of the form m/m, for every nonzero integer m.

(iv) Consider the fractions $-2/4$ and $3/(-6)$. Obviously, 12 is a common denominator of 4 and 6. We write $-2/4$ and $3/(-6)$ with denominator 12. Thus,

$$\frac{-2 \cdot 3}{4 \cdot 3} = \frac{-6}{12} \text{ and } \frac{3 \cdot (-2)}{-6 \cdot (-2)} = \frac{-6}{12}.$$

When written with the same denominator 12, the fractions $-2/4$ and $3/(-6)$ become equal. Consequently, $-2/4 \sim 3/(-6)$.

Remark. One consequence of the definition of equivalence is that if we have a fraction a/b with $b < 0$, there is an equivalent fraction, namely $(-a)/(-b)$, with positive denominator. For we have

$$\frac{a}{b} \sim \frac{a}{b} \cdot \frac{-1}{-1} = \frac{-a}{-b}.$$

and $-b > 0$.

It follows that if the issue at hand involves only equivalence, *we may replace any fraction with an equivalent fraction with a positive denominator.* For example, in our study of the least common denominator in Section 3.3 we were justified in assuming that all denominators are positive. For, if necessary, we can replace a given fraction by an equivalent fraction, as described above, since the sign of numerators does not play a role.

4.2. The Cross Product Criterion for Equivalence

We have discussed the definition of equivalence and worked out some examples. Let us now introduce a very simple criterion for equivalence. Suppose that the fractions a/b and c/d are equivalent. This means that there are nonzero integers h and k so that

$$ah = ck \text{ and } bh = dk.$$

It follows that

$$ah \cdot dk = ck \cdot bh.$$

By associativity and commutativity of multiplication in $\mathbb{Z}$, we have

$$adhk = bchk.$$

Since $hk \neq 0$, by Property (MC), multiplicative cancellation, in $\mathbb{Z}$, we may cancel hk, to obtain the result

$$ad = bc.$$

Thus, we have shown that *if two fractions a/b and c/d are equivalent, then $ad = bc$.*

Next, we want to establish the converse, namely, that if $ad = bc$, then the two fractions a/b and c/d are equivalent. So, let us suppose that $ad = bc$. Then, if we set $h = d$ and $k = b$, by applying the properties of multiplication in $\mathbb{Z}$, we have

$$\frac{ad}{bd} = \frac{bc}{bd} = \frac{cb}{db}$$

and it follows, by definition of equivalence, that $a/b \sim c/d$. Thus, we have found a simple test for equivalence.

If a/b and c/d are fractions, then we have the

Cross Product Criterion for Equivalence of Two Fractions.

$$\frac{\mathbf{a}}{\mathbf{b}} \sim \frac{\mathbf{c}}{\mathbf{d}} \quad \textbf{precisely when} \quad \mathbf{ad = bc}.$$

For fractions a/b and c/d, each of the products ad and bc is known as *a cross product.* Our discussion above demonstrates that the fractions a/b and c/d

are equivalent precisely when their cross products are equal. Thus, the cross product equality $ad = bc$ acquires significance because of the concept of equivalence of fractions.

The cross product criterion makes it a simple matter to check fractions for equivalence. For example, to ascertain whether the fractions $234/585$ and $2/5$ are equivalent, we compute the cross product $234 \cdot 5 = 1170$, and the cross product $585 \cdot 2 = 1170$. Since both cross products are the same, the fractions are equivalent.

Remark. If the fraction a/b has $b < 0$, the cross product criterion gives us another way to show that
$$\frac{a}{b} \sim \frac{-a}{-b},$$
where $-b > 0$.

*

Seminar Exercise.
(i) Find two fractions that are equivalent to $4/5$.
(ii) Find two fractions that are equivalent to $6/1$.
(iii) Find two fractions that are equivalent to $1/(-82)$.
(iv) Let a/b be a fraction and let k be any positive integer. Show that
$$\frac{ka}{kb} \sim \frac{a}{b}.$$

*

Comments on the Seminar Exercise.
(i) For example, $4/5$, $8/10$, $12/15$, and $180/225$ are fractions equivalent to $4/5$. In fact, for any fraction a/b, we have $a/b \sim a/b$.
(ii) For example, $6/1$ and $12/2$ are equivalent to $6/1$. Can you show that $6/1$ is the only fraction with denominator 1 that is equivalent to $6/1$?
(iii) For example, the fractions $1/(-82)$, $(-1)/82$ and $(-2)/164$ are equivalent to $1/(-82)$.
(iv) That $ka/kb \sim a/b$ follows immediately from the cross product criterion.

*

We agreed earlier to identify an integer n with the fraction of the form $n/1$. As one of our first examples in this section we wrote down all fractions equivalent to $1 = 1/1$. Let us now describe all fractions equivalent to $0/1$. By the cross product criterion, we see that $0/2$, $0/5$, $0/(-7)$ are all equivalent to $0/1$. In fact, for any nonzero integer k, the fraction $0/k$ is equivalent to $0/1$ because $0 \cdot 1 = 0 \cdot k = 0$, by the rule for multiplication by 0 found in Section 5 of Seminar 1.

Next, let us ascertain if any fractions other than these can be equivalent to $0/1$. Suppose that $a/b \sim 0/1$. It follows from the cross product criterion that $a \cdot 1 = b \cdot 0$. When we apply the rules for multiplying integers by the multiplicative identity and by the additive identity, respectively, we have $a = 0$. Thus, the fractions equivalent to $0/1$ are precisely the fractions of the form $0/k$, for some nonzero integer k. We say that *a fraction is zero* if it has the form $0/k$ for some nonzero integer k.

4.3. Properties of Equivalence

What properties does the relation of equivalence have? (Two of the three properties defined below, namely reflexivity and transitivity, should be familiar to you from our study of divisibility of integers in Seminar 2.)

Equivalence of fractions is **reflexive,** that is, $a/b \sim a/b$, for every fraction a/b. By commutativity of multiplication in $\mathbb{Z}$, $ab = ba$. Thus, $a/b \sim a/b$ follows from the cross product criterion. (See part (iv) of the Seminar Exercise in Section 4.2.)

Equivalence of fractions is also **symmetric,** that is, the relation $a/b \sim c/d$ implies that $c/d \sim a/b$, for all fractions a/b and c/d. For the hypothesis $a/b \sim c/d$ implies, by the cross product criterion, that $ad = bc$. By commutativity of mutiplication in $\mathbb{Z}$, we have $cb = da$, so $c/d \sim a/b$, again by the cross product criterion.

Finally, we show that equivalence of fractions is **transitive,** which, in this setting means that if a/b, c/d and e/f are fractions, and if $a/b \sim c/d$ and $c/d \sim e/f$, then $a/b \sim e/f$. We translate the two hypotheses into the cross product equations. The hypotheses $a/b \sim c/d$ and $c/d \sim e/f$ imply that

$$ad = bc \quad \text{and} \quad cf = de.$$

We must show that $af = be$. We multiply the first equation above on both sides by f and the second equation above on both sides by b:

$$(ad)f = (bc)f \quad \text{and} \quad (cf)b = (de)b.$$

By commutativity and associativity of multiplication in $\mathbb{Z}$, and by the transitivity of equality, it follows that

$$(af)d = (be)d.$$

By the Property (MC), multiplicative cancellation, in $\mathbb{Z}$, we may cancel the nonzero integer d in this equation. The result is $af = be$, as required.

4.4. Equivalence Classes

We turn now to a very intriguing idea. For every fraction a/b, we form the collection of all of the fractions equivalent to a/b. For example, the collection or class of all fractions equivalent to $2/3$ is this:

$$\left\{ \frac{2}{3}, \frac{-2}{-3}, \frac{4}{6}, \frac{-4}{-6}, \frac{6}{9}, \frac{-6}{-9}, \frac{8}{12}, \cdots \right\},$$

and, for a second example, the collection or class of all fractions equivalent to $-12/8$ is this:

$$\left\{ \frac{-3}{2}, \frac{3}{-2}, \frac{-6}{4}, \frac{6}{-4}, \frac{-9}{6}, \frac{9}{-6}, \frac{-12}{8}, \frac{12}{-8}, \frac{-24}{16}, \frac{24}{-16}, \frac{-36}{24}, \frac{36}{-24}, \frac{-48}{32}, \cdots \right\},$$

∗

Seminar Exercise. Use set notation as in the example above when you describe equivalence classes.

(i) Describe the collection of all fractions equivalent to $3/5$.
(ii) Describe the collection of all fractions equivalent to $(-15)/7$.
(iii) Describe the collection of all fractions equivalent to $72/64$.
(iv) For each collection above, find the fraction in the collection that has smallest positive denominator. What relationship does this fraction have to all of the others in that collection?

∗

Comments on the Seminar Exercise.

(i) The collection of all fractions equivalent to $3/5$ is

$$\left\{ \frac{3}{5}, \frac{-3}{-5}, \frac{6}{10}, \frac{-6}{-10}, \frac{9}{15}, \frac{-9}{-15}, \cdots \right\}.$$

(ii) The collection of all fractions equivalent to $(-15)/7$ is

$$\left\{ \frac{15}{-7}, \frac{-15}{7}, \frac{30}{-14}, \frac{-30}{14}, \frac{45}{-21}, \frac{-45}{21}, \cdots \right\}.$$

(iii) The collection of all fractions equivalent to $72/64$ is

$$\left\{ \frac{9}{8}, \frac{-9}{-8}, \frac{18}{16}, \frac{-18}{-16}, \frac{27}{24}, \frac{-27}{-24}, \cdots, \frac{63}{56}, \frac{-63}{-56}, \frac{72}{64}, \frac{-72}{64}, \frac{81}{72}, \cdots \right\}.$$

(iv) The fraction with smallest positive denominator in the collection in (i) is $3/5$, in (ii) is $(-15)/7$. and in (iii) is $9/8$. Every fraction in each of these collections is obtained by multiplying the fraction with smallest positive denominator in its collection by fractions of the form k/k, where k is a nonzero integer. For example, in the class in (i),

$$\frac{-3}{-5} = \frac{3}{5} \cdot \frac{-1}{-1},$$

in the class in (ii),

$$\frac{-30}{14} = \frac{-15}{7} \cdot \frac{2}{2},$$

in the class in (iii),

$$\frac{72}{64} = \frac{9}{8} \cdot \frac{8}{8}.$$

∗

The collection of all fractions equivalent to a fraction a/b is called the *equivalence class of a/b*.

It follows from the Well-Ordering Property that in each equivalence class there is a fraction having *the smallest positive denominator*. We say that *a fraction is in lowest terms* if it is the fraction in its equivalence class with smallest positive denominator. Consequently, every fraction is equivalent to a fraction in lowest terms. The fraction in lowest terms in its equivalence class is chosen as *the representative of its equivalence class*. In the Seminar Exercise above, The three fractions $3/5$, $(-15)/7$ and $9/8$ are the fractions in lowest terms in their equivalence classes, and, as such, they are chosen to represent their equivalence classes.

※

Seminar Exercise. The following fractions are in lowest terms:

$$\frac{1}{1}, \frac{1}{2}, \frac{7}{10}, \frac{-4}{23}.$$

The following fractions are not in lowest terms:

$$\frac{2}{56}, \frac{-4}{8}, \frac{78}{-13}, \frac{279841}{529}.$$

For each of the fractions not in lowest terms above, find an equivalent fraction that is in lowest terms.

※

Comments on the Seminar Exercise.

$$\frac{2}{56} = \frac{1}{28} \cdot \frac{2}{2},$$

so $2/56 \sim 1/28$. Moreover, $1/28$ is in lowest terms. For if $e/f \sim 1/28$, then by the cross product criterion, $f = 28e$. Consequently, 28 is the smallest positive denominator of all fractions in the equivalence class of $2/56$. Similar arguments show that $-1/2$ is the fraction in lowest terms in the equivalence class of $-4/8$; that $-6/1$ is the fraction in lowest terms in the equivalence class of $78/(-13)$; and that $529/1$ is the fraction in lowest terms in the equivalence class of $279841/529$, because $279841 = (529)^2$.

※

Seminar 9

Properties of Multiplication of Fractions

1. Reducing a Fraction to Lowest Terms

The important observation that begins this seminar connects the definition of lowest terms to the more familiar concept of fractions with numerator and denominator having no factors in common, other than ± 1. We will refer to these results often, so it is crucial that you spend some time absorbing their meaning. Once we have worked through the observation, we will have a quick test for a fraction to be in lowest terms. We recall the definition of "lowest terms." A fraction a/b is in lowest terms means that a/b is the fraction with least positive denominator in its equivalence class. (See Seminar 8, Section 5.4.) For the observation, we assume that the fraction a/b under discussion has positive numerator and denominator. The Corollary that follows demonstrates how to handle the other cases.

OBSERVATION 1.1. *Let a and b be positive integers.*

(i) *If a and b are relatively prime, i.e., $\gcd(a,b) = 1$, then the fraction a/b is in lowest terms.*

(ii) *If a/b is in lowest terms, then a and b are relatively prime.*

(iii) *If a/b is in lowest terms and if $c/d \sim a/b$, then there is a nonzero integer k such that $c = ka$ and $d = kb$. Thus, it follows that if a and b are relatively prime, then every fraction c/d in the equivalence class of a/b has the form*

$$\frac{c}{d} = \frac{a}{b} \cdot \frac{k}{k} = \frac{ak}{bk},$$

where k is a nonzero integer.

We verify the results of the observation. Note that if $b = 1$, then all parts of the observation are clear, so we take $b > 1$.

(i) Assume that a and b are relatively prime. Suppose that c/d is in the equivalence class of a/b and $d > 0$. We must show that $b \leq d$. By the cross product criterion, we have $ad = bc$. This equation implies that b divides the product ad. Since a and b are relatively prime, by Observation 2.2 in Seminar 4, we have b divides d. Thus, we have $b \leq d$, by Property (CDD) in Seminar 2, Section 6.

(ii) Assume that a/b is in lowest terms. Suppose that q is a positive integer that divides both a and b. We must show that $q = 1$. We have $a = cq$ and $b = dq$. Consequently, $a/b = (cq)/(dq) = (c/d)(q/q)$, with $c > 0$ and $d > 0$, and $c/d \sim a/b$. This puts c/d in the same equivalence class as a/b, so $b \leq d$. But $b = dq$, so $q = 1$.

(iii) Assume that a/b is in lowest terms, so that, by (ii), a and b are relatively prime. Suppose $c/d \sim a/b$.

Case 1. We assume first that c and d are positive. By the cross product criterion, $ad = bc$. Consequently, a divides the product bc. Since $\gcd(a, b) = 1$, it follows, by Observation 2.2 in Seminar 4, that a divides c, and $c = ka$, for some positive integer k. For similar reasons, b divides d, and $d = hb$, for some positive integer h. But then we have $kab = hab$, and it follows, by Property (MC) in $\mathbb{Z}$, that $h = k$. Hence, $c/d = ak/bk$.

Case 2. If, on the other hand, both c and d are negative, then $c/d \sim (-c)/(-d)$, where $-c > 0$ and $-d > 0$. By *Case 1*, there is a positive integer k such that

$$\frac{-c}{-d} = \frac{ka}{kb}.$$

Thus,

$$\frac{c}{d} = \frac{(-k)a}{(-k)b}.$$

COROLLARY 1.2. *Assume that a and b are positive integers as in the observation and that a/b is in lowest terms. Then*

(i) *The fraction $(-a)/b$ is in lowest terms.*
(ii) *The fraction $(-a)/b$ is the fraction in lowest terms equivalent to $a/(-b)$.*
(iii) *The fraction a/b is the fraction in lowest terms equivalent to $(-a)/(-b)$.*

We leave the verification of the corollary to the interested reader.

The instruction: "Reduce the fraction to lowest terms" simply means, "Replace the fraction by the fraction in lowest terms in its equivalence class."

Steps to Reduce a Fraction c/d to Lowest Terms

(i) Suppose that c and d are positive integers. Follow the steps below to reduce the fraction c/d to lowest terms.
Step (i) Compute $\gcd(c, d) = k$.
Step (ii) Write $c = ka$ and $d = kb$.
Step (iii) Cancel the $k's$:

$$\frac{c}{d} = \frac{ka}{kb} \sim \frac{a}{b}.$$

(ii) Suppose that c and d are not both positive integers.
Suppose, for example, that c is positive and that $d = -e$, where e is positive. First, we reduce the fraction c/e to lowest terms a/b, where $a > 0$ and $b > 0$, by following the steps above. We then have $c/d \sim -a/b$, by part (ii) of Corollary 1.2. (The case c negative and d positive, and the case both c and d negative are handled in an analogous fashion.)

Examples.
1. Suppose we are asked to reduce the fraction $70/(-84)$ to lowest terms. Since the denominator -84 is negative, we reduce the fraction $70/84$ to lowest terms first. To do that, we follow steps (i) - (iii).
(i) $\gcd(70, 84) = 14$
(ii) $70 = 14 \cdot 5$, and $84 = 14 \cdot 6$
(iii) Cancel the 14 in the numerator and the 14 in the denominator:

$$\frac{70}{84} = \frac{14 \cdot 5}{14 \cdot 6} \sim \frac{5}{6}.$$

Thus, $5/6$ is the fraction in lowest terms equivalent to $70/84$. It follows from part (ii) of Corollary 1.2 that $-5/6$ is the fraction in lowest terms equivalent to $70/(-84)$.

2. Reduce the fraction $(-495)/166$ to lowest terms. We apply the Euclidean Algorithm to 495 and 166 :

$$495 = 2 \cdot 166 + 163$$
$$166 = 1 \cdot 163 + 3$$
$$163 = 54 \cdot 3 + 1$$
$$3 = 3 \cdot 1 + 0$$

Thus, $\gcd(495, 166) = 1$ and, by part (i) of Corollary 1.2, the fraction $(-495)/166$ *is* in lowest terms.

✳

Seminar Exercise. Reduce the following fractions to lowest terms:

$$\frac{63}{6}, \frac{-25}{-15}, \frac{-210}{52}, \frac{153}{-117}.$$

✳

Comments on the Seminar Exercise.
(i) We have $\gcd(63, 6) = 3$, and $63 = 3 \cdot 21$ and $6 = 3 \cdot 2$. It follows that

$$\frac{63}{6} = \frac{3 \cdot 21}{3 \cdot 2} \sim \frac{21}{2}$$

and $21/2$ is in lowest terms.

(ii) It is obvious by inspection that $25/15 \sim 5/3$ which is in lowest terms. Thus, by part (iii) of Corollary 1.2, it follows that $5/3$ is the fraction in lowest terms equivalent to $(-25)/(-15)$.

(iii) First, reduce the fraction $210/52$ to lowest terms. We have $\gcd(210, 52)$ $= 2$, and $210 = 2 \cdot 105$ and $52 = 2 \cdot 26$. Consequently, $105/26$ is the fraction in lowest terms equivalent to $210/52$. By part (i) of Corollary 1.2, it follows that $-105/26$ is the fraction in lowest terms equivalent to $-210/52$.

(iv) We have $\gcd(153, 117) = 9$, and $153 = 9 \cdot 17$ and $117 = 9 \cdot 13$. Consequently, $153/117 \sim 17/13$ and $17/13$ is in lowest terms. By part (ii) of Corollary 1.2, we have $(-17)/13$ is the fraction in lowest terms equivalent to $153/(-117)$.

✳

Remark. We must be ready to recognize that sometimes fractions not in lowest terms are used to represent an equivalence class. For example, on a gasoline pump in a service station, the amount of gas is ordinarily listed in tenths of a gallon. Consequently, we might read

$$7 \frac{5}{10}$$

gallons on the pump, where $5/10$ is not in lowest terms.

✳

Seminar/Classroom Discussion. What are some other examples where fractions not in lowest terms are used to represent their equivalence classes?

✳

2. Multiplication and Equivalence

Equivalence is a fundamental relation among fractions. It is essential that the basic operations of multiplication and addition of fractions respect this relationship. We will discuss addition later, but we have defined and used multiplication of fractions extensively in Seminar 8 as well as this seminar, so we explain now what it means for multiplication of fractions to respect equivalence. An analogy is the comparison of equivalence as a gear that is vital to the design of the fraction machine, and of multiplication as another gear essential to the machine's operation. The gears must mesh or the machine breaks.

Suppose a/b, a'/b', c/d and c'/d' are fractions with

$$\frac{a}{b} \sim \frac{a'}{b'} \quad \text{and} \quad \frac{c}{d} \sim \frac{c'}{d'}.$$

We want to show that

$$\frac{a}{b} \cdot \frac{c}{d} \sim \frac{a'}{b'} \cdot \frac{c'}{d'}.$$

To put it differently, if a/b and a'/b' are in one equivalence class, and c/d and c'/d' are in a second equivalence class, possibly the same as the first, then we want to demonstrate that the products

$$\frac{a}{b} \cdot \frac{c}{d} \quad \text{and} \quad \frac{a'}{b'} \cdot \frac{c'}{d'}$$

are in the same equivalence class. Once this is done, we are assured that the equivalence gear and the multiplication gear work together.

First, we look at an example. Consider the fractions 2/7 and 3/8. We have $2/7 \sim 6/21$ and $3/8 \sim 12/32$. We must show that

$$\frac{2}{7} \cdot \frac{3}{8} \sim \frac{6}{21} \cdot \frac{12}{32}$$

We have

$$\frac{2}{7} \cdot \frac{3}{8} = \frac{6}{56} \quad \text{and} \quad \frac{6}{21} \cdot \frac{12}{32} = \frac{72}{672}$$

To demonstrate that the products are equivalent, we apply the cross product criterion. We have $6 \cdot 672 = 4032$ and $56 \cdot 72 = 4032$. Consequently, the product fractions are equivalent, as desired.

We record the general case as the next observation.

OBSERVATION 2.1. *If a/b, a'/b', c/d and c'/d' are fractions with*

$$\frac{a}{b} \sim \frac{a'}{b'} \quad \text{and} \quad \frac{c}{d} \sim \frac{c'}{d'}$$

then,

$$\frac{a}{b} \cdot \frac{c}{d} \sim \frac{a'}{b'} \cdot \frac{c'}{d'}.$$

We use the cross product criterion to translate the hypothesis

$$\frac{a}{b} \sim \frac{a'}{b'} \quad \text{and} \quad \frac{c}{d} \sim \frac{c'}{d'}$$

into two simple equations of cross products:

$$ab' = ba' \quad \text{and} \quad cd' = dc'.$$

Using the properties of multiplication in $\mathbb{Z}$, we have

$$acb'd' = bda'c',$$

and it follows, again by the cross product criterion, that

$$\frac{ac}{bd} \sim \frac{a'c'}{b'd'},$$

that is

$$\frac{a}{b} \cdot \frac{c}{d} \sim \frac{a'}{b'} \cdot \frac{c'}{d'}.$$

In other words, the observation states that "multiplication is a class operation," that is, if we multiply any of the fractions in the equivalence class of a/b by a fraction in the equivalence class of c/d, we obtain a fraction in the equivalence class of the product ac/bd. Moreover, every fraction in the equivalence class of the product ac/bd is equivalent to the product of a fraction in the equivalence class of a/b and a fraction in the equivalence class of c/d.

3. Properties of Multiplication of Fractions

The multiplication rule for fractions is based on the multiplication rule for the integers, and, as noted in Seminar 8, Section 4.1, it follows directly from the properties of multiplication of integers that multiplication of fractions has the properties of closure, associativity and commutativity. Thus, for fractions a/b, c/d and e/f, multiplication has the following three properties.

(M1) Closure The product of two fractions is a fraction.

(M2) Associativity

$$\left[\frac{a}{b} \cdot \frac{c}{d}\right] \cdot \frac{e}{f} = \frac{a}{b} \cdot \left[\frac{c}{d} \cdot \frac{e}{f}\right]$$

(M3) Commutativity

$$\frac{a}{b} \cdot \frac{c}{d} = \frac{c}{d} \cdot \frac{a}{b}.$$

When we discussed closure, associativity and commutativity in Seminar 8, Section 4.1, we did not address the existence of a multiplicative identity or the existence of multiplicative inverses for nonzero fractions. We will do so now. To treat these topics we must rely on equivalent fractions.

(M4) Multiplicative Identity The fraction $1/1$, which we identify with the integer 1, is the fraction in lowest terms satisfying

$$\frac{a}{b} \cdot \frac{1}{1} = \frac{a \cdot 1}{b \cdot 1} = \frac{a}{b}.$$

Thus, there is a multiplicative identity for fractions. Recall, however, that if k is a nonzero integer, then

$$\frac{k}{k} \sim \frac{1}{1}$$

and

$$\frac{a}{b} \cdot \frac{k}{k} = \frac{ak}{bk} \sim \frac{a}{b}.$$

It follows that there are many fractions that act as a multiplicative identity in terms of equivalence, but all of them are in the same equivalence class. In other words, if we multiply a fraction a/b by a fraction k/k, then

$$\frac{a}{b} \cdot \frac{k}{k} \sim \frac{a}{b}.$$

However,

$$\frac{a}{b} \cdot \frac{k}{k} = \frac{a}{b}$$

precisely when $k = 1$.

Recall that a fraction is nonzero exactly when its numerator is nonzero.
(M5) Multiplicative Inverses for Nonzero Fractions For each nonzero

fraction a/b, there is a fraction, namely the fraction b/a, called the *reciprocal* of a/b, such that

$$\frac{a}{b} \cdot \frac{b}{a} \sim \frac{1}{1}.$$

First of all, note that b/a is a fraction because $a \neq 0$, since a/b is nonzero. Next, we look at the product of a/b and b/a :

$$\frac{a}{b} \cdot \frac{b}{a} = \frac{ab}{ba} = \frac{ab}{ab}.$$

We have just seen that fractions of the form ab/ab act as an identity for multiplication of fractions. Thus,

$$\frac{a}{b} \cdot \frac{b}{a} = \frac{ab}{ab} \sim \frac{1}{1}.$$

This illustrates the fact that when we multiply fractions we are really multiplying equivalence classes.

4. Equality and Equivalence

For the sake of both clarity and convenience, from now on we will *replace the equivalence symbol, $\sim$, with the equality symbol, $=$* . This means that the equation $c/d = a/b$ denotes that the fraction c/d is in the equivalence class of a/b. Thus, when we write $2/4 = 1/2$ and $56/24 = 7/3$, the reader must understand that the fractions $2/4$ and $1/2$ are not equal; they are equivalent, and the fractions $56/24$ and $7/3$ are not equal; they are equivalent. Other examples to take note of are the following.

(i) For any nonzero integer k, when we write

$$\frac{k}{k} = \frac{1}{1} = 1,$$

the first "$=$" means equivalence and the second "$=$" is identification with the integer 1.

(ii) For any nonzero integer k, when we write

$$\frac{0}{k} = \frac{0}{1} = 0,$$

the first "$=$" means equivalence and the second "$=$" is identification with the integer 0.

We have shown that multiplication of fractions is a "class operation," and we will also show that addition of fractions is also a "class operation." Consequently, this change of notation does not disrupt the mechanics of the arithmetic of fractions.

5. Fractions and Mixed Numbers

Up to now, in our treatment of fractions there has been no attempt to relate the size of the numerator to that of the denominator. We have simply stated that a fraction is an expression a/b, where a and b are integers and $b \neq 0$. Thus, the expressions

$$\frac{4}{7}, \frac{31}{5}, \frac{-5}{11}, \text{ and } \frac{57}{-23}$$

are all fractions. Common terminology frequently assigns the name "proper fraction" to a fraction of the form a/b with $a > 0$, $b > 0$ and $a < b$. All other fractions must then be in the undesirable category of "improper fractions." Of the four fractions above, only 4/7 is deemed "proper." Our attitude is that no fraction is improper as long as its denominator is nonzero.

For measuring purposes it is useful to express a fraction as a *mixed number*, that is, as the sum of an integer and a fraction with positive numerator, positive denominator, and with the numerator less than the denominator. This allows us to estimate quantities to the nearest unit. Consider the sizes S, M and L of women's hats. Size M corresponds to head measurements of $21\frac{7}{8} - 22\frac{1}{4}$ in, so a head measurement to the nearest eighth is needed to decide whether M is the correct size.

We explain how to write a fraction in this way. If we have a fraction a/b with $a > 0$, $b > 0$ and $a > b$, we apply the division algorithm and write

$$a = bq + r, \text{ with } q > 0 \text{ and } 0 \leq r < b.$$

Then we multiply both sides of this equation by the fraction $1/b$ and apply the distributive property, which will be discussed in Seminar 10, Section 5, to obtain:

$$\frac{a}{b} = q + \frac{r}{b}, \text{ where } q > 0 \text{ and } 0 \leq r < b.$$

When we express $q + r/b$ as a mixed number, we write

$$\frac{a}{b} = q\frac{r}{b}.$$

For example, the fraction 21/16 expressed as a mixed number is $1\frac{5}{16}$ and is read as "one and five sixteenths." The word "and" substitutes for the + in the equation

$$\frac{21}{16} = 1 + \frac{5}{16}.$$

Note. The concept of expressing fractions as mixed numbers can be extended to fractions with negative numerator and positive denominator by extending the division algorithm to include the case where the dividend is negative.

※

Seminar Exercise. Write each of the following fractions as a mixed number:

$$\frac{15}{14}, \ \frac{67}{17}, \ \frac{59}{28}.$$

Comments on the Seminar Exercise.
We write the fraction $15/14$ in the form $15 = 1 \cdot 14 + 1$, and multiply both sides of the equation by $1/14$, to obtain the equation

$$\frac{15}{14} = 1 + \frac{1}{14}.$$

Thus, the fraction $15/14$ can be written as the mixed number $1\frac{1}{14}$.

We write $67 = 3 \cdot 17 + 16$, and multiply both sides of this equation by $1/17$ to obtain the equation:

$$\frac{67}{17} = 3 + \frac{16}{17}.$$

Thus, we write the fraction $67/17$ as the mixed number $3\frac{16}{17}$, and we observe that $67/17$ is a bit less than the integer 4.

We write $59 = 2 \cdot 28 + 3$, and multiply both sides of this equation by $1/28$ to obtain the equation:

$$\frac{59}{28} = 2 + \frac{3}{28}.$$

We write $2\frac{3}{28}$ and observe that the fraction $59/28$ is slightly larger than the integer 2.

✳

6. Division of Fractions

Let us review what we know about division in the integers. Recall that to divide an integer a by an integer b means that there is an integer c such that $bc = a$, or, in other words, there is an integer solution $x = c$ to the equation $bx = a$. We saw that such a solution does not always exist, and this led to the introduction of the division algorithm for integers.

In contrast, division of any fraction by a nonzero fraction can always be defined and we will do so now. To divide a fraction c/d by a nonzero fraction a/b means to find a fraction, briefly named x and called the *quotient* of c/d by a/b, that satisfies the equation

$$\frac{a}{b} x = \frac{c}{d}.$$

What makes division of fractions possible is the fact that every nonzero fraction has a multiplicative inverse, something we know does not hold in $\mathbb{Z}$. We illustrate how to divide one fraction by a nonzero fraction with an

example. To divide $7/16$ by $5/9$ means to find a fraction x that is a solution
to the equation:
$$\frac{5}{9}x = \frac{7}{16}.$$
The nonzero fraction $5/9$ has a multiplicative inverse $9/5$ so we have, by our
agreement at the beginning of Section 4 to replace $\sim$ by $=$,
$$\frac{9}{5} \cdot \frac{5}{9} = \frac{45}{45} = 1.$$
We multiply both sides of the equation $(5/9)x = 7/16$, by $9/5$:
$$\left(\frac{9}{5}\right)\left(\frac{5}{9}\right)x = \left(\frac{9}{5}\right)\left(\frac{7}{16}\right),$$
and simplify to obtain
$$1 \cdot x = \left(\frac{9}{5}\right)\left(\frac{7}{16}\right).$$
Thus,
$$x = \frac{63}{80}.$$
We see that if we want to divide $7/16$ by $5/9$, we multiply $7/16$ by $9/5$, the
multiplicative inverse (i.e., the reciprocal) of $5/9$.

It is just as straightforward to derive the definition of *division* of any
fraction c/d by a nonzero fraction a/b. To solve the equation
$$\frac{a}{b}x = \frac{c}{d},$$
i.e., to find the quotient x of c/d by a/b, we multiply both sides of the
equation above by the reciprocal b/a of a/b :
$$\left(\frac{b}{a}\right)\left(\frac{a}{b}\right)x = \left(\frac{b}{a}\right)\left(\frac{c}{d}\right).$$
Since $(b/a)(a/b) = 1$, we obtain
$$x = \frac{b}{a} \cdot \frac{c}{d}.$$
Thus,
$$\frac{c}{d} \div \frac{a}{b} = \frac{b}{a} \cdot \frac{c}{d} = \frac{bc}{ad}.$$
Remark. By Property (M3), the Commutative Rule for Multiplication of
fractions, we may write the solution x above in the form
$$\frac{c}{d} \div \frac{a}{b} = \frac{c}{d} \cdot \frac{b}{a} = \frac{cb}{da}.$$
It is sometimes more useful to think of the fractions in this way because it
preserves, on the right side, the order found on the left.

Recall the old adage "Ours is not to reason why, just invert and multiply." The discussion above explains the "reason why" we "just invert and multiply." One of the purposes of these seminars is to "reason why."

✳

Seminar Exercise. Compute the quotient in each of the following division problems

$$\text{(i)} \quad \frac{4}{5} \div \frac{9}{11}$$

$$\text{(ii)}(ii) \quad \frac{3}{7} \div \frac{2}{5}$$

$$\text{(iii)} \quad \frac{9}{26} \div 13$$

$$\text{(iv)} \quad \frac{-2}{7} \div \frac{10}{3}$$

$$\text{(v)} \quad \frac{13}{230} \div \frac{22}{83}$$

✳

Comments on the Seminar Exercise.

(i)
$$\frac{4}{5} \div \frac{9}{11} = \frac{11}{9} \cdot \frac{4}{5} = \frac{4}{5} \cdot \frac{11}{9} = \frac{44}{45}$$

(ii)
$$\frac{3}{7} \div \frac{2}{5} = \frac{3}{7} \cdot \frac{5}{2} = \frac{15}{14}.$$

(iii)
$$\frac{9}{26} \div 13 = \frac{9}{26} \div \frac{13}{1} = \frac{9}{26} \cdot \frac{1}{13} = \frac{9}{338}.$$

(iv)
$$\frac{-2}{7} \div \frac{10}{3} = \frac{-2}{7} \cdot \frac{3}{10} = \frac{-6}{70}$$

(v)
$$\frac{13}{230} \div \frac{22}{83} = \frac{13}{230} \cdot \frac{83}{22} = \frac{1079}{5060}$$

✳

7. Word Problems with Fractions

Next, let us try some word problems with fractions. These problems illustrate how useful mixed numbers are for estimating the sizes of the quantities involved in the problems. It is important, and often helpful, to express these quantities using the units designated in the problems. Although calculations can be made without carrying the units along, the final answer must include the units. We adopt the usual abbreviations for the units of time and measurement. We express rates in fractional form, for example, we write miles per hour as miles/hour.

Examples.
1. How many crepe paper streamers of length $1\frac{2}{5}$ ft can be cut from a $15\frac{1}{2}$ ft roll?

In essence, the problem asks, "How many seven-fifths are there in thirty-one halves?" To find out, we divide $31/2$ by $7/5$:

$$\frac{31}{2} \div \frac{7}{5} = \frac{31}{2} \cdot \frac{5}{7} = \frac{155}{14} = 11\frac{1}{14}.$$

Consequently, 11 streamers of length $1\frac{2}{5}$ ft can be cut from the $15\frac{1}{2}$ roll.

2. Our supermarket installed new "do it yourself" scanners for the Express Check Out lanes. The average time for a customer to check out in these lanes is $2\frac{5}{6}$ minutes. How many customers using one of these lanes can check out in $\frac{1}{2}$ hour?

To respect the units in the problem, we change $1/2$ hour to 30 minutes. As in the first problem, we are asked, in essence, to find how many two and five-sixths there are in thirty. Thus, we divide $30/1$ by $17/6$:

$$\frac{30}{1} \div \frac{17}{6} = \frac{30}{1} \cdot \frac{6}{17} = \frac{180}{17} = 10\frac{10}{17}.$$

Thus, on average, 10 customers can check out in 30 minutes.

Note that in each of the first two problems, the fractional part of the result was discarded because, given the problem's scenario, it did not make sense to include it.

Observe that the supermarket checkout problem above studied the number of customers checking out in a certain time. In other words, it looked at the time rate of change of customers checking out. The next two problems involve the rate of change of distance with respect to time, often called *speed*. In such problems, the distance a person or animal or car or plane, etc. travels is equal to the rate at which it travels multiplied by the time of travel:

$$\text{distance} = \text{rate} \cdot \text{time}$$

abbreviated

$$d = r \cdot t.$$

3. The giraffe is known for its speed. The maximum speed for a giraffe is approximately $6/10$ mi/min (miles per minute). How long, in minutes, does it take for a giraffe moving at maximum speed to run $2\frac{1}{2}$ miles?

We look, for a moment, at the units involved in the problem. The distance is in miles, speed is in miles per minute and the time is in minutes, so symbolically, the units in the equation $d = r \cdot t$:

$$\text{miles} \;=\; \frac{\text{mi}}{\text{min}} \cdot \text{minutes}$$

are correct "mathematically."

We are given d and r and are required to solve the equation $d = r \cdot t$ for t :

$$\frac{5}{2} = \frac{6}{10} \cdot t.$$

We must divide the distance by the speed. To find the time for the giraffe to travel $2\frac{1}{2}$ miles, we divide $5/2$ by the speed $6/10$ mi/min. We have

$$\frac{5}{2} \div \frac{6}{10} = t$$
$$\frac{5}{2} \cdot \frac{10}{6} = t$$
$$\frac{50}{12} = t$$
$$4\frac{1}{6} = t$$

Thus, the giraffe runs $2\frac{1}{2}$ miles in $4\frac{1}{6}$ minutes.

4. The maximum speed of a hedgehog is approximately $27/4$ feet per second. How long does it take for a hedgehog, running at maximum speed, to travel 500 feet? Broadly, compare with the giraffe.

The units are different here, perhaps in recognition of the hedgehog's size. We use the equation $d = r \cdot t$, where d is the distance traveled in feet, r is the speed in feet per second and t is the time in seconds. To calculate the time it takes the hedgehog to cover 500 feet, we divide 500 by the fraction $27/4$:

$$500 \div \frac{27}{4} = 500 \cdot \frac{4}{27} = \frac{2000}{27} = 74\frac{2}{27} \text{ sec}$$

Thus, the hedgehog runs 500 feet in $74\frac{2}{27}$ seconds. How does the hedgehog's speed compare to that of the giraffe? The hedgehog runs a little less than $1/10$ mile in about 70 seconds. The giraffe runs $1/10$ mile in approximately 10 seconds. The hedgehog is very fast for its size.

✳

Seminar/Classroom Activity. Discuss and solve the following problems.

  (i) To prepare for a race, an athlete plans to run 8 miles a day. If the path around the lake in the park is $1\frac{2}{7}$ miles long, how many laps on the path must the athlete run each day to accomplish the goal?

(ii) The road between two ponds is 2/3 mile long. A land tortoise, the slowest member of the turtle/tortoise family, walks, on average, 1/5 mi/hr. How long would it take a tortoise walking at this speed to make the trek on the road from one pond to the other?

(iii) The school garden is $23\frac{1}{3}$ sq ft. The students will be planting tomato seedlings this year. Each seedling requires $1\frac{3}{4}$ sq ft of space. How many tomato seedlings can be planted in the garden?

✳

Comments on the Seminar Activity.

(i) We need to find how many one and two-sevenths there are in eight. So we divide 8 by $\frac{9}{7}$:

$$8 \div \frac{9}{7} = 8 \cdot \frac{7}{9} = \frac{56}{9} = 6\frac{2}{9} \text{ laps.}$$

Thus, the runner must do 7 laps on the path around the lake to cover the desired 8 miles.

(ii) This is a speed problem, so we use the fact that the distance d covered is equal to the rate r of travel multiplied by the time t. To compute the time for the tortoise to walk 2/3 mile, we divide d by r :

$$\frac{2}{3} \div \frac{1}{5} = \frac{2}{3} \cdot \frac{5}{1} = 3\frac{1}{3} \text{ hours.}$$

It will take the tortoise $3\frac{1}{3}$ hours to make the trek.

(iii) For this problem, we seek how many seven-fourths there are in seventy thirds:

$$\frac{70}{3} \div \frac{7}{4} = \frac{70}{3} \cdot \frac{4}{7} = \frac{40}{3} = 13\frac{1}{3}.$$

The students can plant 13 tomato seedlings.

✳

Seminar 10

Addition of Fractions

In this seminar, we assume that the denominators of all fractions are positive.

1. Addition of Fractions with the Same Denominator

As in the classroom, our discussion of addition begins with fractions having the same denominator. There is no difficulty motivating addition of fractions with a common denominator. Children recognize that it is as natural as the familiar addition of integers.

We cut two circles out of two differently colored papers and cut one of the circles into quarters. (We use the second circle as our model of the whole circle.) By the third and fourth grades, students understand that one quarter of the circle added to another quarter of the circle is two fourths of the circle. Some may recognize that the sum is one half of the circle. Expressed mathematically, we have

$$\frac{1}{4} + \frac{1}{4} = \frac{2}{4} = \frac{1}{2}.$$

Furthermore, young children see that if we add another quarter to the half of the circle that is the sum of two quarter cicles, we have three quarters of the circle:

$$\frac{1}{2} + \frac{1}{4} = \frac{1}{4} + \frac{1}{4} + \frac{1}{4} = \frac{3}{4}.$$

The rule for adding two fractions a/d and c/d with the same denominator d bears that out:

$$\frac{a}{d} + \frac{c}{d} = \frac{a+c}{d},$$

that is, "add the numerators, using integer addition, and keep the denominator."

Addition of fractions with the same denominator, or, as we say, with a *common denominator*, is a straightforward extension of addition of integers.

2. The Rule for Addition of Fractions

When students are comfortable with addition of fractions with the same denominator, we describe how to convert a problem of adding fractions with different denominators into one where the denominators are the same. For example, to explain how to add the fractions 1/4 and 1/6, we cut two new circles out of two differently colored papers and cut one of the circles in half. If one of the half-circles is cut in half, then each of the two pieces is one quarter-circle. If the other half-circle is cut into thirds, then, by the "part of a part" rule, each of the pieces is one sixth of a circle, for $(1/3)(1/2) = 1/6$.

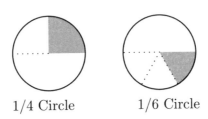

1/4 Circle   1/6 Circle

To answer the question of what part of the whole circle is one quarter-circle added to one sixth of the circle, we describe how the quarter-circle and the one-sixth-circle can be cut up further into smaller sections of the same size. The quarter-circle can be cut into thirds, resulting in three pieces where each one is, by the part of a part rule, equal to one twelfth of the circle: $(1/3)(1/4) = 1/12$. The one-sixth-circle can be cut in half, resulting in two pieces where each one is, by the part of a part rule, also equal to one twelfth of the circle. Thus, we are now dealing with fractions with a common denominator. We add the three twelfths obtained from the quarter-circle to the two twelfths obtained from the one-sixth-circle to obtain 5/12 of the circle:

$$\frac{1}{4} + \frac{1}{6} = \frac{3}{12} + \frac{2}{12} = \frac{5}{12}.$$

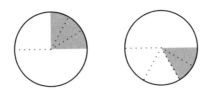

5/12 Circle

Since we know how to add two fractions if the denominators are the same, we consider the addition of two fractions a/b and c/d, where $b \neq d$. Our strategy is to show that each of the two fractions can be written with *a common denominator equal to the product of the denominators of the two fractions.* We explain how we do it with an example first. Let us find the sum of $1/3$ and $1/7$. Since for any nonzero integer k, the fraction $k/k = 1$, by applying the properties of multiplication of fractions, we may write

$$\frac{1}{3} = \frac{1}{3} \cdot \frac{7}{7} = \frac{1 \cdot 7}{3 \cdot 7} = \frac{7}{21}.$$

What we are doing is multiplying the *first* fraction by 1, where 1 is represented by the fraction with numerator and denominator equal to the denominator of the *second* fraction. (Note that all of the "equal signs" in the equation above stand for equivalence.) For the same reasons, we may write

$$\frac{1}{7} = \frac{1}{7} \cdot \frac{3}{3} = \frac{1 \cdot 3}{7 \cdot 3} = \frac{3}{21}.$$

What we are doing here is multiplying the *second* fraction by 1 where 1 is represented by the fraction with numerator and denominator equal to the denominator of the *first* fraction. Thus we have found two fractions, namely $7/21$ and $3/21$, *with a common denominator* that are equivalent to the given fractions $1/3$ and $1/7$, respectively. We know how to add such fractions: add the numerators, using integer addition, and keep the denominator:

$$\frac{7}{21} + \frac{3}{21} = \frac{7+3}{21} = \frac{10}{21}.$$

The sum $10/21$ is in lowest terms. In examples where the sum is not in lowest terms, cancel, as necessary, after adding.

We write down a sequence of steps for adding any two fractions. We call it the "sure fire" method because the steps are the *same* for *every* problem. Suppose we are asked to find the sum of the fractions a/b and c/d.

Sure Fire Method for Finding the Sum of a/b and c/d.

(i) Multiply the fraction a/b by the fraction d/d :

$$\frac{a}{b} \cdot \frac{d}{d} = \frac{ad}{bd}$$

to obtain the equivalent fraction ad/bd.

(ii) Multiply the fraction c/d by the fraction b/b :

$$\frac{c}{d} \cdot \frac{b}{b} = \frac{cb}{db} = \frac{bc}{bd}$$

to obtain the equivalent fraction $cb/db = bc/bd$.

(iii) Now we have a problem of finding the sum of fractions with a common denominator:

$$\frac{a}{b} + \frac{c}{d} = \frac{ad}{bd} + \frac{bc}{bd}.$$

We apply the rule for adding fractions with a common denominator, that is, add the numerators and keep the denominator.

$$\frac{a}{b} + \frac{c}{d} = \frac{ad}{bd} + \frac{bc}{bd} = \frac{ad+bc}{bd}.$$

That's all there is to it!

To find the sum of two fractions with different denominators, the sure fire method converts the two fractions to two equivalent fractions with a common denominator, and so, to find the sum, we add the numerators and keep the common denominator. Observe that the common denominator in the sure fire method is the product of the denominators of the summands.

Here are some examples.

(i)

$$\frac{2}{3} + \frac{1}{5} = \frac{2}{3} \cdot \frac{5}{5} + \frac{1}{5} \cdot \frac{3}{3}$$
$$= \frac{2 \cdot 5}{3 \cdot 5} + \frac{1 \cdot 3}{5 \cdot 3} = \frac{2 \cdot 5 + 1 \cdot 3}{3 \cdot 5}$$
$$= \frac{10+3}{15} = \frac{13}{15}.$$

(ii)

$$\frac{3}{7} + \frac{8}{5} = \frac{3}{7} \cdot \frac{5}{5} + \frac{8}{5} \cdot \frac{7}{7}$$
$$= \frac{15}{35} + \frac{56}{35} = \frac{71}{35}.$$

(iii)

$$\frac{1}{6} + \frac{-3}{10} = \frac{10+(-18)}{60} = \frac{-8}{60}.$$

The sum $-8/60$ is not in lowest terms. Both the numerator and the denominator are divisible by 4. Consequently, we may write

$$\frac{-8}{60} = \frac{-2 \cdot 4}{15 \cdot 4} = \frac{-2}{15},$$

which is in lowest terms.

✳

Seminar Exercise. Use the sure fire method to find the following sums.

$$\text{(i)} \quad \frac{1}{11} + \frac{3}{10} \qquad \text{(ii)} \quad \frac{251}{26} + \frac{-53}{4} \qquad \text{(iii)} \quad \frac{47}{52} + \frac{91}{89}$$

✳

Comments on the Seminar Exercise.

(i)

$$\begin{aligned}
\frac{1}{11} + \frac{3}{10} &= \frac{1}{11} \cdot \frac{10}{10} + \frac{3}{10} \cdot \frac{11}{11} \\
&= \frac{10}{110} + \frac{33}{110} = \frac{43}{110}.
\end{aligned}$$

(ii)

$$\begin{aligned}
\frac{251}{26} + \frac{-53}{4} &= \frac{251}{26} \cdot \frac{4}{4} + \frac{-53}{4} \cdot \frac{26}{26} \\
&= \frac{1004}{104} + \frac{-1378}{104} \\
&= \frac{-374}{104} = \frac{-187}{52} \cdot \frac{2}{2} = \frac{-187}{52}
\end{aligned}$$

(iii)

$$\begin{aligned}
\frac{47}{52} + \frac{91}{89} &= \frac{47}{52} \cdot \frac{89}{89} + \frac{91}{89} \cdot \frac{52}{52} \\
&= \frac{4183}{4628} + \frac{4732}{4628} = \frac{8915}{4628}.
\end{aligned}$$

✳

3. Comparison of Methods for Addition of Fractions

Most of us learned to add fractions by means of the least common denominator. Why is it taught this way? The main argument is that the result is in lowest terms. But that may not be the case; see Parts (i) and (ii) of the Seminar Exercise below. The process of finding the least common denominator has several steps, at any one of which students are at risk for mistakes. We think that, with the exception of a few cases where the least common denominator is obvious, students should be taught to use the sure fire method and, if required, reduce the result to lowest terms.

Consider the problem of finding the sum of the fractions 1/8 and 3/22. We find the sum by the least common denominator method and by the sure fire method.

Problem. Find

$$\frac{1}{8} + \frac{3}{22}.$$

(1) Least Common Denominator Method.
Before we begin, recall that the least common denominator, abbreviated lcd, of two fractions a/b and c/d is the least common multiple of their denominators, and that

$$\operatorname{lcd}\left(\frac{a}{b}, \frac{c}{d}\right) = \operatorname{lcm}[b, d] = \frac{bd}{\gcd(b, d)},$$

see Seminar 8, Section 4.3.

(i) To apply the formula above, we first compute $\gcd(8, 22)$ by the Euclidean Algorithm:

$$22 = 2 \cdot 8 + 6$$
$$8 = 1 \cdot 6 + 2$$
$$6 = 3 \cdot 2 + 0.$$

Thus, $\gcd(8, 22) = 2$. By the formula above, $\operatorname{lcm}[b, d] = (8 \cdot 22)/2 = 88$. (We could have calculated $\operatorname{lcm}[8, 22]$ directly from the prime factorizations, see Seminar 5, Section 3.)

(ii) The next step is to write each fraction as a fraction with denominator 88. We observe that $88 = 8 \cdot 11$. Thus,

$$\frac{1}{8} = \frac{1}{8} \cdot \frac{11}{11} = \frac{11}{88}.$$

Similarly, $88 = 22 \cdot 4$, so

$$\frac{3}{22} = \frac{3}{22} \cdot \frac{4}{4} = \frac{12}{88}.$$

(iii) We find the sum of the equivalent fractions with least common denominator by adding the numerators and keeping the denominator.

$$\frac{1}{8} + \frac{3}{22} = \frac{11}{88} + \frac{12}{88} = \frac{23}{88}.$$

It is Steps (i) and (ii) that cause some students to stumble. These steps are avoided in the next method.

(2) The Sure Fire Method.
Both fractions are written with denominator equal to the product $8 \cdot 22$ of the denominators.

(i) We multiply $1/8$ by the fraction $22/22$:

$$\frac{1}{8} \cdot \frac{22}{22} = \frac{22}{176}.$$

(ii) We multiply $3/22$ by the fraction $8/8$:

$$\frac{3}{22} \cdot \frac{8}{8} = \frac{24}{176}.$$

(iii) We add the equivalent fractions with common denominator:

$$\frac{1}{8} + \frac{3}{22} = \frac{22}{176} + \frac{24}{176} = \frac{46}{176}.$$

(iv) The numerator and denominator are both divisible by 2, so, if the sum is required to be in lowest terms, we cancel the factor 2 in both the numerator and denominator:

$$\frac{1}{8} + \frac{3}{22} = \frac{46}{176} = \frac{23}{88}$$

✳

Seminar Exercise. Calculate the following sums using the method of finding the least common denominator.

(i)
$$\frac{3}{10} + \frac{5}{14}$$

Is the result in lowest terms?

(ii)
$$\frac{5}{42} + \frac{7}{60}$$

Is the result in lowest terms?

(iii) Calculate both sums above by the sure fire method.

✳

Comments on the Seminar Exercise.

(i) We calculate the least common denominator of the fractions using the formula

$$\text{lcd}\left(\frac{a}{b}, \frac{c}{d}\right) = \text{lcm}[b, d] = \frac{bd}{\gcd(b, d)}.$$

First, we use the Euclidean Algorithm to compute $\gcd(10, 14)$.

$$14 = 1 \cdot 10 + 4$$
$$10 = 2 \cdot 4 + 2$$
$$4 = 2 \cdot 2 + 0.$$

Thus, $\gcd(10, 14) = 2$. (We could also have observed that $\text{lcm}[10, 14] = 2 \cdot 5 \cdot 7 = 70$ directly from the prime factorizations, see Seminar 5, Section 3.) Therefore,

$$\text{lcd}\left(\frac{3}{10}, \frac{5}{14}\right) = \text{lcm}[10, 14] = \frac{140}{2} = 70.$$

Since $70 = 10 \cdot 7$, and $70 = 14 \cdot 5$, we have

$$\frac{3}{10} = \frac{3}{10} \cdot \frac{7}{7} = \frac{21}{70}$$

and

$$\frac{5}{14} = \frac{5}{14} \cdot \frac{5}{5} = \frac{25}{70}.$$

We have two fractions with a common denominator and we add:

$$\frac{3}{10} + \frac{5}{14} = \frac{21}{70} + \frac{25}{70} = \frac{46}{70}.$$

The sum is not in lowest terms. Its numerator and denominator are divisible by 2. We cancel the 2, and obtain

$$\frac{46}{70} = \frac{23 \cdot 2}{35 \cdot 2} = \frac{23}{35}.$$

The fraction $23/35$ is in lowest terms.

(ii) First, we find $\gcd(42, 60)$ by the Euclidean Algorithm.

$$60 = 1(42) + 18$$
$$42 = 2(18) + 6$$
$$18 = 3(6) + 0.$$

Thus, $\gcd(42, 60) = 6$. By the formula above,

$$\mathrm{lcd}\left(\frac{5}{42}, \frac{7}{60}\right) = \mathrm{lcm}[42, 60] = \frac{42 \cdot 60}{6} = 420.$$

We have $420 = 42 \cdot 10 = 60 \cdot 7$. Consequently,

$$\frac{5}{42} + \frac{7}{60} = \frac{50}{420} + \frac{49}{420} = \frac{99}{420} = \frac{33}{140} \cdot \frac{3}{3} = \frac{33}{140}.$$

The fraction $99/420$ is not in lowest terms. The numerator and denominator are divisible by 3. When we reduce to lowest terms, we have

$$\frac{5}{42} + \frac{7}{60} = \frac{33}{140}.$$

(iii) We do the problem in (i) by the sure fire method.

$$\frac{3}{10} \cdot \frac{14}{14} + \frac{5}{14} \cdot \frac{10}{10} = \frac{42}{140} + \frac{50}{140} = \frac{92}{140}.$$

The sum is not is lowest terms as both numerator and denominator are divisible by 4 :

$$\frac{92}{140} = \frac{23}{35} \cdot \frac{4}{4} = \frac{23}{35}.$$

We do the problem in (ii) by the sure fire method.

$$\frac{5}{42} \cdot \frac{60}{60} + \frac{7}{60} \cdot \frac{42}{42} = \frac{300 + 294}{2520} = \frac{594}{2520}$$

The numerator and denominator of the result above are divisible by 18, so

$$\frac{5}{42} + \frac{7}{60} = \frac{594}{2520} = \frac{33}{140} \cdot \frac{18}{18} = \frac{33}{140}.$$

Even though the summand fractions in both of the problems in this Seminar Exercise are in lowest terms, observe that when computed by the least common denominator method, neither result is in lowest terms.

$$*$$

Remarks.
(i) It is important to note that it is not necessary to find the *least* common denominator, *any* common denominator will do. Observe that the sure fire method uses the common denominator that is the product of the two denominators.

(ii) In the case where b and d are relatively prime, the least common denominator of the fractions a/b and c/d is equal to the product bd of the denominators. Consequently, in this case, both methods of calculating the sum $a/b + c/d$ are the same.

4. The Properties of Addition of Fractions

The properties of addition of fractions are the same as the properties for addition of integers and are derived from them. Keep in mind that although we have replaced the $\sim$ symbol by $=$, equivalence is omnipresent in discussions of fractions.

(A1) Closure. If a/b and c/d are fractions, then

$$\frac{a}{b} + \frac{c}{d}$$

is a fraction.

(A2) Associative Rule. If a/b, c/d and e/f are fractions, then

$$\left(\frac{a}{b} + \frac{c}{d}\right) + \frac{e}{f} = \frac{a}{b} + \left(\frac{c}{d} + \frac{e}{f}\right).$$

(A3) Commutative Rule. For fractions a/b and c/d,

$$\frac{a}{b} + \frac{c}{d} = \frac{c}{d} + \frac{a}{b}.$$

(A4) Additive Identity. The fraction $0/1$, which we identify with the integer 0 is the fraction in lowest terms satisfying

$$\frac{a}{b} + \frac{0}{1} = \frac{0}{1} + \frac{a}{b} = \frac{a}{b},$$

for all fractions a/b.

(A5) Additive Inverse. A fraction a/b has an additive inverse, $(-a)/b$. This means that

$$\frac{a}{b} + \frac{-a}{b} = \frac{0}{1}.$$

Consider Property (A1). To verify closure for addition, we must show that $(ad+bc)/bd$ is a fraction, i.e., that $ad+bc$ is an integer and that bd is a nonzero integer. The first statement follows from closure of addition and multiplication in $\mathbb{Z}$, and the second statement follows from Property (NZ) in Seminar 2. Observe that the denominator of the sum and of the product of the two fractions a/b and c/d is the same: bd.

✳

Seminar Exercise. Verify Properties (A3), (A4) and (A5). Remember addition of fractions is based on addition and multiplication in $\mathbb{Z}$. Use the properties of these operations in $\mathbb{Z}$ as you verify (A3), (A4) and (A5) for fractions. If you enjoy a challenge, verify Property (A2).

✳

Comments on the Seminar Exercise.
For Property (A3), we apply the rule for addition of fractions:
$$\frac{a}{b}+\frac{c}{d}=\frac{ad+bc}{bd},$$
and then we use commutativity of both addition and multiplication of integers to write
$$\frac{ad+bc}{bd}=\frac{cb+da}{db}.$$
Since
$$\frac{cb+da}{db}=\frac{c}{d}+\frac{a}{b},$$
by definition of addition of fractions, it follows that
$$\frac{a}{b}+\frac{c}{d}=\frac{c}{d}+\frac{a}{b}.$$
For Property (A4), it is sufficient to show that
$$\frac{a}{b}+\frac{0}{1}=\frac{a}{b},$$
for all fractions a/b. We have
$$\frac{a}{b}+\frac{0}{1}=\frac{a\cdot1+0\cdot b}{b\cdot1}.$$
But,
$$\frac{a\cdot1+0\cdot b}{b\cdot1}=\frac{a+0}{b}=\frac{a}{b},$$
since 1 is the multiplicative identity in $\mathbb{Z}$ and 0 is the additive identity in $\mathbb{Z}$.

To verify Property (A5), we apply the definition of addition of fractions with the same denominator:
$$\frac{a}{b}+\frac{(-a)}{b}=\frac{a+(-a)}{b}.$$

Next, we apply the properties of additive inverse for integers:

$$\frac{a + (-a)}{b} = \frac{0}{b}.$$

We know that the fraction $0/b$ is equivalent to the fraction $0/1$ which we identify with the integer 0. Thus, it follows that

$$\frac{a}{b} + \frac{(-a)}{b} = \frac{0}{1} = 0.$$

✳

Since, as we have just established, every fraction a/b has an additive inverse $(-a)/b$, we may introduce the symbol

$$-\frac{a}{b}$$

(also written $-(a/b)$) to designate the additive inverse. Although the expression $-(a/b)$ is not a fraction, it is used to represent the equivalence class of the fraction $(-a)/b$. The equality

$$-\frac{a}{b} = \frac{-a}{b}$$

simply indicates that $-(a/b)$ is being used to represent the equivalence class of $(-a)/b$.

The fact that every fraction has an additive inverse makes it possible to define subtraction of fractions. The operation of *subtraction of fractions* is defined in the same way as subtraction of integers. The difference $a/b - (c/d)$ of the fractions a/b and c/d is defined as follows.

$$\frac{a}{b} - \frac{c}{d} = \frac{a}{b} + \left(-\frac{c}{d}\right).$$

Consequently,

$$\frac{a}{b} - \frac{c}{d} = \frac{a}{b} + \frac{-c}{d} = \frac{a}{b} \cdot \frac{d}{d} + \frac{-c}{d} \cdot \frac{b}{b} = \frac{ad - bc}{bd}.$$

For instance,

$$\frac{3}{5} - \frac{2}{9} = \frac{3}{5} + \frac{-2}{9} = \frac{27 - 10}{45} = \frac{17}{45}.$$

✳

Seminar Exercise. Compute the following:

$$(i)\ \frac{2}{7} - \frac{1}{9} \quad (ii)\ -\frac{10}{3} + \frac{15}{8}.$$

✳

Comments on the Seminar Exercise. For (i), we have

$$\frac{2}{7} - \frac{1}{9} = \frac{2}{7} + \frac{-1}{9}$$
$$= \frac{2}{7} \cdot \frac{9}{9} + \frac{-1}{9} \cdot \frac{7}{7}$$
$$= \frac{18}{63} + \frac{-7}{63} = \frac{18 - 7}{63} = \frac{11}{63}.$$

For (ii), we have

$$-\frac{10}{3} + \frac{15}{8} = \frac{-10}{3} + \frac{15}{8}$$
$$= \frac{-10}{3} \cdot \frac{8}{8} + \frac{15}{8} \cdot \frac{3}{3}$$
$$= \frac{-80}{24} + \frac{45}{24} = \frac{-80 + 45}{24} = \frac{-35}{24}.$$

✳

5. The Distributive Property

The property that asserts that multiplication distributes over addition in the set of fractions depends upon the fact that multiplication distributes over addition in $\mathbb{Z}$. We establish this property by applying the distributive property in $\mathbb{Z}$.

(D) Distributive Property. For fractions, a/b, c/d and e/f,

$$\frac{a}{b} \cdot \left(\frac{c}{d} + \frac{e}{f} \right) = \frac{a}{b} \cdot \frac{c}{d} + \frac{a}{b} \cdot \frac{e}{f}$$

We demonstrate distributivity step by step. We alert you again to the fact that, although we have replaced $\sim$ by $=$, equivalence is everywhere.

$$\frac{a}{b} \cdot \left(\frac{c}{d} + \frac{e}{f} \right) = \frac{a}{b} \cdot \left(\frac{cf + de}{df} \right),$$

by the definition of addition of fractions. By definition of multiplication of fractions, we have

$$\frac{a}{b} \cdot \left(\frac{cf + de}{df} \right) = \frac{(a)(cf + de)}{b(df)}$$

By the distributive property in $\mathbb{Z}$, we may write

$$\frac{(a)(cf + de)}{b(df)} = \frac{(a)(cf) + (a)(de)}{b(df)}$$

We pause for a moment here to consider our goal. We are trying to verify the distributive property for multiplication and addition of fractions. To reach this goal, we must show that

$$\frac{(a)(cf) + (a)(de)}{b(df)} = \frac{a}{b} \cdot \frac{c}{d} + \frac{a}{b} \cdot \frac{e}{f}$$

In the final step, we are going to pull apart the sum

$$\frac{(a)(cf) + (a)(de)}{b(df)}$$

into two fractions that are equivalent to the fractions ac/bd and ae/bf.

Now we continue our work. By associativity in $\mathbb{Z}$ it follows that,

$$\frac{(a)(cf) + (a)(de)}{b(df)} = \frac{(ac)(f) + (ad)(e)}{b(df)}.$$

We use associativity and commutativity of multiplication and the definition of addition in reverse to write

$$\frac{(ac)(f) + (ad)(e)}{b(df)} = \frac{(ac)(f)}{(bd)(f)} + \frac{(ae)(d)}{(bf)(d)}.$$

Observe that the first fraction on the right above is equivalent to ac/bd and the second to ae/bf. Thus, we may write

$$\frac{(ac)(f)}{(bd)(f)} + \frac{(ae)(d)}{(bf)(d)} = \frac{ac}{bd} + \frac{ae}{bf}.$$

By the definition of multiplication of fractions, we have

$$\frac{ac}{bd} + \frac{ae}{bf} = \frac{a}{b} \cdot \frac{c}{d} + \frac{a}{b} \cdot \frac{e}{f},$$

as required.

By the commutative property (A3) for fractions, it also follows that

$$\left(\frac{a}{b} + \frac{c}{d}\right) \cdot \frac{e}{f} = \frac{a}{b} \cdot \frac{e}{f} + \frac{c}{d} \cdot \frac{e}{f}.$$

6. Addition and Equivalence

If a/b and c/d are fractions, then, by definition of addition,

$$\frac{a}{b} + \frac{c}{d} = \frac{ad + bc}{bd}.$$

We want to show, as we did in Seminar 9, Section 2 for multiplication, that addition is a "class operation." In other words, we want to show that if a'/b' and c'/d' are fractions with $a'/b' \sim a/b$ and $c'/d' \sim c/d$, then

$$\frac{ad + bc}{bd} \sim \frac{a'd' + b'c'}{b'd'}$$

For example, if we add $1/2$ and $1/5$, then the sum is

$$\frac{1}{2} + \frac{1}{5} = \frac{7}{10}.$$

We have $3/6 \sim 1/2$ and $2/10 \sim 1/5$. Let us show that $3/6 + 2/10 \sim 7/10$. The sum of $3/6$ and $2/10$ is

$$\frac{3}{6} + \frac{2}{10} = \frac{3 \cdot 10 + 2 \cdot 6}{6 \cdot 10} = \frac{42}{60}.$$

This fraction is equivalent to $7/10$, for

$$\frac{42}{60} = \frac{6 \cdot 7}{6 \cdot 10} \sim \frac{7}{10}.$$

It is true that no matter which fraction we take in the equivalence class of $1/2$ and which fraction we take in the equivalence class of $1/5$, the sum is equivalent to $7/10$. If you enjoy algebraic manipulation, you will enjoy the seminar exercise below which demonstrates that addition of fractions is a class operation.

✳

Seminar Exercise. Let a/b, a'/b', c/d and c'/d' be fractions. Suppose that

$$\frac{a}{b} \sim \frac{a'}{b'} \quad \text{and} \quad \frac{c}{d} \sim \frac{c'}{d'},$$

Show that

$$\frac{a}{b} + \frac{c}{d} \sim \frac{a'}{b'} + \frac{c'}{d'}.$$

✳

Comments on the Seminar Exercise. By the cross product criterion, we have $ab' = ba'$ and that $cd' = dc'$. We must show that

$$\frac{ad + bc}{bd} \sim \frac{a'd' + b'c'}{b'd'}.$$

We want to verify this equivalence by the cross product criterion, so we must demonstrate that

$$(ad + bc)(b'd') = (bd)(a'd' + b'c').$$

This is a good exercise in careful substitution and manipulation of addition and multiplication of integers. Here we go. We start with the left side of the equality displayed above. Many properties of arithmetic in the integers are invoked. We use the distributive property and associativity first:

$$(ad + bc)(b'd') = adb'd' + bcb'd'.$$

Then we apply commutativity to write

$$adb'd' + bcb'd' = ab'dd' + bb'cd'.$$

We substitute ba' for ab', and dc' for cd':

$$ab'dd' + bb'cd' = ba'dd' + bb'dc.'$$

Finally, we regroup to obtain the expression on the right side of the equation we sought to verify.

$$ba'dd' + bb'dc' = bd(a'd' + b'c').$$

✳

In this section, we have shown that addition depends only on the equivalence classes of the fractions being added. In Seminar 9, Section 2, we demonstrated that multiplication depends only on the equivalence classes of the fractions being multiplied. Taken together, these results establish the fact that multiplication and addition of fractions are *independent of the particular representative chosen from each class.* The formal mathematical statement of this fact is that multiplication and addition of fractions are "well defined."

Seminar 11

The Decimal Expansion of a Fraction

1. Definitions of Decimal Fraction, Decimal Expansion and Decimal

In Seminar 9, Section 5, we discussed mixed numbers and observed that, given a fraction fraction c/d, with $c > d > 0$, we may apply the division algorithm

$$c = dq + r, \quad \text{where } 0 \leq r < d,$$

to obtain

$$\frac{c}{d} = q + \frac{r}{d}.$$

We write this sum in the standard notation for mixed numbers:

$$\frac{c}{d} = q\frac{r}{d},$$

with juxtaposition replacing $+$.

If, in fact, $r = 0$, then $c/d = q$ is an integer. It follows, as a result, that we may focus on the decimal expansion of fractions a/b with $0 < a < b$. We will see in Seminar 12, Section 3, that the fractions a/b with $0 < a < b$ are precisely the fractions with $0 < a/b < 1$.

Unless we explicitly say otherwise, we assume for the remainder of this seminar that **all fractions a/b satisfy $0 < a < b$**. In this seminar, we intend to write a/b as a sum of fractions that have denominators equal to powers of 10. A fraction with numerator less than denominator and with denominator a power of 10 is called a *decimal fraction*. For example, $5/10$ and $3/100$ are decimal fractions. The most basic decimal fractions are unit decimal fractions, such as,

$$\frac{1}{10}, \frac{1}{10^2}, \frac{1}{10^3}, \frac{1}{10^4}, \ldots, \frac{1}{10^n}, \ldots$$

We will see that *every* fraction a/b, with $0 < a/b < 1$, can be written as a sum of decimal fractions, each having numerator equal to a decimal digit, that is, one of the integers $0, 1, \ldots, 9$. Such a sum is called a *decimal expansion* of the fraction. As we will see, for example,

$$\frac{43}{80} = \frac{5}{10} + \frac{3}{10^2} + \frac{7}{10^3} + \frac{5}{10^4} = \frac{5375}{10^4}.$$

We write

$$\frac{5375}{10^4} = 0.5375.$$

The expression 0.5375 is is called the *decimal (or base* 10*) representation of the decimal fraction* $5375/10^4$, or *decimal*, for short. A decimal is a concise way of writing a sum of decimal fractions.

The "period," around which we place decimal digits, is called the *decimal point*. The first place to the right of the decimal point is called the *tenths* place; the second place to the right of the decimal point is called the *hundredths* place; the third place to the right of the decimal point is called the *thousandths* place and the fourth place to the right of the decimal point is called the *ten thousandths* place. The 0 to the left of the decimal point, in the *units* place, indicates that the fraction is between 0 and 1. The digit occupying each place is called the *place value*. We can read off the place value from the decimal expansion or from the decimal. Reading from the decimal expansion, the value of the 10^{-j} ths place is the numerator of the decimal fraction in the expansion that has denominator 10^j. Reading from the decimal, the 10^{-j} ths place value is the integer occupying the 10^{-j} ths place. For the decimal

$$\frac{43}{80} = \frac{5}{10} + \frac{3}{10^2} + \frac{7}{10^3} + \frac{5}{10^4} = 0.5375,$$

the tenths place has value 5, the hundredths place has value 3, the thousandths place has value 7 and the ten thousandths place has value 5.

More generally, we will see that every fraction a/b can be written in the form

$$\frac{a}{b} = \frac{q_1}{10} + \frac{q_2}{10^2} + \frac{q_3}{10^3} + \frac{q_4}{10^4} + \cdots = 0.q_1 q_2 q_3 q_4 \ldots,$$

where the q_i are decimal digits. Just as for the example above, the 0 to the left of the decimal point in the units place signifies that $0 < a/b < 1$, the tenths place has value q_1, the hundredths place has value q_2, the thousandths place has value q_3, etc. In general, we say that the 10^{-j} ths place has value q_j. Consequently, when we show how to construct the decimal expansion, we must find the value for each decimal place. This we will do in the next section.

Suppose a fraction c/d has $c > d > 0$, and we write it as a mixed number,

$$\frac{c}{d} = q\frac{r}{d},$$

where q is a positive integer, and r is an integer with $0 \le r < d$. Then the decimal expansion of c/d is the integer q written in base 10, and the decimal expansion of r/d, where $0 < r/d < 1$. Thus, a number such as $328.749 = 328 + 0.749$ has value 3 in the hundreds place, value 2 in the tens place, value 8 in the units place, value 7 in the tenths place, value 4 in the hundredths place and value 9 in the thousandths place.

There are two types of decimals (and decimal expansions): those that "stop" such as 0.5375, and those that do "not stop" such as the familiar decimal expansion of

$$\frac{1}{3} = \frac{3}{10} + \frac{3}{10^2} + \frac{3}{10^3} + \cdots + \frac{3}{10^k} + \cdots = 0.333\ldots .$$

Those that "stop" are called terminating decimals, and those that don't "stop" are called nonterminating decimals. A precise definition follows.

Suppose the decimal expansion of the fraction a/b is

$$\frac{a}{b} = \frac{q_1}{10} + \frac{q_2}{10^2} + \frac{q_3}{10^3} + \cdots ,$$

where the place values q_i are decimal digits, that is, integers from 0 to 9. If there is a positive integer k, such that $q_k \neq 0$, but $q_j = 0$, for all $j > k$, then this decimal expansion is called a *terminating decimal expansion*. Its decimal expansion is written

$$\frac{a}{b} = \frac{q_1}{10} + \frac{q_2}{10^2} + \frac{q_3}{10^3} + \cdots + \frac{q_k}{10^k}.$$

Its decimal, called a *terminating decimal*, is written $0.q_1q_2 \cdots q_k$.

If, on the other hand, for every positive integer k, there is an integer $j > k$, such that $q_j \neq 0$, then the decimal expansion is said to be a *nonterminating decimal expansion* with decimal expansion

$$\frac{a}{b} = \frac{q_1}{10} + \frac{q_2}{10^2} + \frac{q_3}{10^3} + \cdots + \frac{q_j}{10^j} + \cdots ,$$

and *nonterminating decimal* $0.q_1q_2q_3\ldots$. Thus, $43/80$ has a terminating decimal expansion, while $1/3$ has a nonterminating decimal expansion. We will study both types of expansions in detail, and show you how to recognize the fractions that have terminating expansions without having to calculate the expansion first.

We close this section with some examples demonstrating that every terminating decimal expansion can be written as a decimal fraction, and conversely that every fraction equivalent to a decimal fraction has a terminating decimal expansion. Consider the decimal expansion

$$\frac{3}{10} + \frac{2}{10^2} + \frac{8}{10^3} = 0.328.$$

The fractions on the left above have common denominator 10^3, so, by the properties of addition of fractions, we have

$$\frac{3}{10} + \frac{2}{10^2} + \frac{8}{10^3} = \frac{3 \cdot 10^2 + 2 \cdot 10 + 8}{10^3} = \frac{300 + 20 + 8}{10^3} = \frac{328}{10^3} = 0.328.$$

Thus the decimal 0.328 is the same as the decimal fraction $328/10^3$. Observe that this decimal fraction is not in lowest terms.

Next, consider the decimal fraction $915/10000 = 915/10^4$. Since 915 can be written $915 = 9 \cdot 10^2 + 1 \cdot 10 + 5$, it follows that

$$\frac{915}{10^4} = \frac{9 \cdot 10^2 + 1 \cdot 10 + 5}{10^4}.$$

By the rules of addition of fractions, we have

$$\frac{9 \cdot 10^2 + 1 \cdot 10 + 5}{10^4} = \frac{9 \cdot 10^2}{10^4} + \frac{1 \cdot 10}{10^4} + \frac{5}{10^4},$$

and, by cancelling powers of 10, we have

$$\frac{915}{10^4} = \frac{9}{10^2} + \frac{1}{10^3} + \frac{5}{10^4} = 0.0915.$$

This example illustrates how a decimal may be written as a decimal fraction and a decimal fraction may be written as a decimal.

2. Constructing the Decimal Expansion of a Fraction

Recall that we are assuming, unless specifically stated otherwise, that every fraction a/b satisfies $0 < a < b$, i.e., satisfies $0 < a/b < 1$. We discuss how to construct the decimal expansion of a/b :

$$\frac{a}{b} = \frac{q_1}{10} + \frac{q_2}{10^2} + \frac{q_3}{10^3} + \cdots,$$

where the q_i are decimal digits, that is, integers from 0 to 9. This procedure is often taught by means of the long division algorithm. Our aim, in this section, is to show you that the decimal expansion can be constructed by repeated applications of the division algorithm. In the Appendix, we explain how each step of the long division algorithm emerges from this process.

To construct the decimal expansion of a/b, we find the decimal place values q_i, step by step, as follows.

Step 1. We apply the division algorithm to $10a$ and b. This furnishes us with integers q_1 and r_1, such that

$$10a = q_1 b + r_1, \quad \text{where } 0 \le r_1 < b.$$

Next, we divide both sides of this equation by $10b$ to obtain

$$\frac{a}{b} = \frac{q_1}{10} + \frac{r_1}{10b}.$$

This gives us a *first approximation* to a/b by a decimal fraction and tells us q_1, the tenths place value in the decimal expansion of a/b. If $r_1 \ne 0$, we proceed to the next step which is to apply the division algorithm to $10r_1$ and b.

Step 2. The division algorithm for $10r_1$ and b yields integers q_2 and r_2, such that

$$10r_1 = q_2 b + r_2, \quad \text{where } 0 \le r_2 < b.$$

Next, we divide both sides of this equation by $10^2 b$ to obtain

$$\frac{r_1}{10b} = \frac{q_2}{10^2} + \frac{r_2}{10^2 b}.$$

When we substitute the expression on the right above for $r_1/10b$ in the equation

$$\frac{a}{b} = \frac{q_1}{10} + \frac{r_1}{10b},$$

we have

$$\frac{a}{b} = \frac{q_1}{10} + \frac{q_2}{10^2} + \frac{r_2}{10^2 b}.$$

This gives us the hundredths place value q_2 in the decimal expansion of a/b. If $r_2 \neq 0$, we proceed to the next step which is to apply the division algorithm to $10r_2$ and b.

Step 3. Applying the division algorithm to divide $10r_2$ by b yields integers q_3 and r_3, such that

$$10r_2 = q_3 b + r_3, \quad \text{where } 0 \leq r_3 < b.$$

Next, we divide both sides of this equation by $10^3 b$ to obtain

$$\frac{r_2}{10^2 b} = \frac{q_3}{10^3} + \frac{r_3}{10^3 b},$$

and, when we substitute the expression on the right above for $r_2/10^2 b$ in the equation

$$\frac{a}{b} = \frac{q_1}{10} + \frac{q_2}{10^2} + \frac{r_2}{10^2 b},$$

we have the thousandths place value q_3 in the decimal expansion of a/b:

$$\frac{a}{b} = \frac{q_1}{10} + \frac{q_2}{10^2} + \frac{q_3}{10^3} + \frac{r_3}{10^3 b}.$$

If, for some k, the remainder $r_k = 0$, but the previous remainder $r_{k-1} \neq 0$, the procedure ends with Step k. The result is that a/b has a terminating decimal expansion:

$$\frac{a}{b} = \frac{q_1}{10} + \frac{q_2}{10^2} + \frac{q_3}{10^3} + \cdots + \frac{q_k}{10^k}.$$

If, on the other hand, for every k, the remainder $r_k \neq 0$, then the procedure continues indefinitely, and the decimal expansion is nonterminating:

$$\frac{a}{b} = \frac{q_1}{10} + \frac{q_2}{10^2} + \frac{q_3}{10^3} + \cdots + \frac{q_j}{10^j} + \cdots.$$

However, note that even in the nonterminating case, at every step j, we have the equality

$$\frac{a}{b} = \frac{q_1}{10} + \frac{q_2}{10^2} + \frac{q_3}{10^3} + \cdots + \frac{q_j}{10^j} + \frac{r_j}{10^j b},$$

which, since $r_j < b$, results in a *decimal approximation to a/b to within $1/10^j$*:

$$\frac{a}{b} - \frac{q_1}{10} - \frac{q_2}{10^2} - \frac{q_3}{10^3} - \cdots - \frac{q_j}{10^j} = \frac{r_j}{10^j b} < \frac{1}{10^j}.$$

Take note that the decimal place values are the quotients q_i that arise in the repeated applications of the division algorithm.

We work out some examples.

Example 1. Let us calculate the decimal expansion of $43/80$ using the method of repeated applications of the division algorithm. We have $a = 43$ and $b = 80$.

Step 1. We apply the division algorithm to $10 \cdot 43$ and 80 :

$$10 \cdot 43 = 5 \cdot 80 + 30,$$

and find that $q_1 = 5$ and $r_1 = 30$. Thus, 5 is the digit in the tenths place. Since $30 = r_1 \neq 0$, we proceed to the next step.

Step 2. We apply the division algorithm to $10 \cdot 30$ and 80 :

$$10 \cdot 30 = 3 \cdot 80 + 60,$$

where $q_2 = 3$ and $r_2 = 60$. The digit in the hundreths place of the decimal expansion of $43/80$ is 3 and $r_2 = 60 \neq 0$. We proceed to the next step.

Step 3. We apply the division algorithm to $10 \cdot 60$ and 80 :

$$10 \cdot 60 = 7 \cdot 80 + 40,$$

where $q_3 = 7$ is the digit in the thousandths place and $r_3 = 40$. Since $r_3 \neq 0$, ordinarily we would proceed to the next step. But note that $40/80000 = 5/10^4$, so we have

$$\frac{43}{80} = \frac{5}{10} + \frac{3}{10^2} + \frac{7}{10^3} + \frac{5}{10^4} = 0.5375$$

Example 2. Next, we calculate the decimal expansion of $5/11$. We have $a = 5$ and $b = 11$.

Step 1. We apply the division algorithm to $10 \cdot 5$ and 11 :

$$10 \cdot 5 = 50 = 4 \cdot 11 + 6.$$

So the digit in the tenths place of the decimal expansion of $5/11$ is $q_1 = 4$ and $r_1 = 6$.

Step 2. We apply the division algorithm to $10 \cdot 6$ and 11 :

$$10 \cdot 6 = 5 \cdot 11 + 5.$$

Thus, the digit in the hundredths place is $q_2 = 5$, and $r_2 = 5$. Observe that $r_2 = 5 = a$, so the procedure will repeat because in the next step we will divide $10a = 50$ by $b = 11$ again and obtain the same result as in Step 1. To convince ourselves of this, we do one more step.

Step 3. We apply the division algorithm to $10 \cdot 5$ and 11 :

$$10 \cdot 5 = 50 = 4 \cdot 11 + 6,$$

where $q_3 = q_1 = 4$, and $r_3 = r_1 = 6$. Consequently, after the second step, the remainders cycle between 5 and 6 and the quotients (place values), between 4 and 5, respectively. Thus, $5/11$ has a nonterminating decimal expansion that repeats after two places:

$$\frac{5}{11} = \frac{4}{10} + \frac{5}{10^2} + \frac{4}{10^3} + \frac{5}{10^4} + \cdots = 0.454545\ldots.$$

This is an example of a nonterminating decimal expansion in which a sequence of place values repeats after a fixed number of steps. In this example the sequence 45 repeats after 2 steps.

Example 3. We find the decimal expansion of the fraction $1/7$.

Step 1. We apply the division algorithm to $10 \cdot 1$ and 7 :

$$10 \cdot 1 = 1 \cdot 7 + 3,$$

where $q_1 = 1$ and $r_1 = 3$.
Step 2. We apply the division algorithm to $10 \cdot 3$ and 7 :

$$10 \cdot 3 = 4 \cdot 7 + 2,$$

where $q_2 = 4$ and $r_2 = 2$.

Step 3. We apply the division algorithm to $10 \cdot 2$ and 7 :

$$10 \cdot 2 = 2 \cdot 7 + 6,$$

where $q_3 = 2$ and $r_3 = 6$.

Step 4.

$$10 \cdot 6 = 60 = 8 \cdot 7 + 4,$$

where $q_4 = 8$ and $r_4 = 4$.

Step 5.

$$10 \cdot 4 = 40 = 5 \cdot 7 + 5,$$

where $q_5 = 5$ and $r_5 = 5$.

Step 6.

$$10 \cdot 5 = 50 = 7 \cdot 7 + 1,$$

where $q_6 = 7$ and $r_6 = 1$.

Finally, we have a repetition! Observe that $r_6 = 1$, the numerator of $1/7$, so the division algorithm process will repeat. The decimal expansion of $1/7$ is

$$\frac{1}{7} = \frac{1}{10} + \frac{4}{10^2} + \frac{2}{10^3} + \frac{8}{10^4} + \frac{5}{10^5} + \frac{7}{10^6} + \frac{1}{10^7} + \frac{4}{10^8} + \cdots = 0.1428571\ldots.$$

It repeats after six steps.

We emphasize the following two **very important** observations.

OBSERVATION 2.1. *It may happen that a place value $q_j = 0$, but $r_j \neq 0$.*

For example, the fraction $1/11$ has nonterminating repeating decimal expansion

$$\frac{1}{11} = \frac{0}{10} + \frac{9}{10^2} + \frac{0}{10^3} + \frac{9}{10^4} + \cdots = 0.090909\ldots.$$

The odd place values are equal to 0 and the even place values are equal to 9. However $r_j \neq 0$, for all j.

OBSERVATION 2.2. *If the decimal expansion of a fraction a/b does not terminate, then there is a sequence of place values of length at most $b-1$, that repeats beginning at the 10^{-b} ths place or earlier.*

To see why this statement is true, note that in each step of the division algorithm procedure for calculating the decimal expansion of a/b, the remainder is one of the integers $0, 1, 2, \ldots, b-1$. It follows that if the remainder is never 0, then, the remainder is one of the integers $1, 2, \ldots, b-1$. (Remember that a is one of these integers since $0 < a < b$.) Thus there are $b-1$ possible remainders. It follows from the pigeonhole principle (see Appendix B) that, by the bth step, or earlier, one of the remainders must repeat or be equal to a. Consequently, as we saw in Examples 2 and 3 above, the place values in the decimal expansion itself will repeat because the quotients that arise in the division algorithm will repeat. We have what is called a *repeating decimal expansion*. The sequence of recurring place values is called the *repetend* and the number of place values in the repetend is called its *length*. Moreover, we have just deduced, from the pigeonhole principle, that the length of the repetend is at most $b-1$. Our examples show that the fraction $5/11$ has repetend 45 of length 2, and the fraction $1/7$ has repetend 142857 which has length 6, the longest possible. A terminating decimal $0.q_1q_2 \cdots q_k$ can be regarded as a repeating decimal $0.q_1q_2 \cdots q_k\,0\,0\,0 \ldots$ with repetend equal to 0.

✳

Seminar Exercise. Use the division algorithm procedure to calculate the decimal expansion of each of the following fractions. If it is a nonterminating decimal expansion, identify the repetend and its length.

$$\frac{3}{8}; \; \frac{1}{9}; \; \frac{5}{6}; \; \frac{9}{16}; \; \frac{4}{13}.$$

✳

Comments on the Seminar Exercise.

For 3/8, we have the following repeated division algorithm steps:

$$10 \cdot 3 = 3 \cdot 8 + 6,$$
$$10 \cdot 6 = 7 \cdot 8 + 4,$$
$$10 \cdot 4 = 5 \cdot 8 + 0.$$

Thus,

$$\frac{3}{8} = 0.375,$$

which we could write, with repetend 0, as $0.375\,0\,0\,\dots$.

For 1/9, we have

$$10 \cdot 1 = 1 \cdot 9 + 1,$$

Thus, $r_1 = 1$, the numerator of 1/9. So the place values repeat immediately. Consequently,

$$\frac{1}{9} = 0.11111\dots .$$

This decimal has repetend 1 of length 1.

For 5/6, we have

$$10 \cdot 5 = 50 = 8 \cdot 6 + 2,$$

$$10 \cdot 2 = 20 = 3 \cdot 6 + 2.$$

Thus, $r_2 = r_1$, but r_1 is not the numerator of 5/6. This decimal expansion is repeating with repetend equal to 3, which, you will observe, is the hundredths place value. Consequently,

$$\frac{5}{6} = 0.83333\dots .$$

For 9/16, we have the repeated division algorithm steps:

$$90 = 5 \cdot 16 + 10,$$
$$100 = 6 \cdot 16 + 4,$$
$$40 = 2 \cdot 16 + 8,$$
$$80 = 5 \cdot 16 + 0.$$

Thus,

$$\frac{9}{16} = 0.5625 = 0.562500000\dots .$$

For 4/13, we have the repeated division algorithm steps:

$$40 = 3 \cdot 13 + 1,$$
$$10 = 0 \cdot 13 + 10,$$
$$100 = 7 \cdot 13 + 9,$$
$$90 = 6 \cdot 13 + 12,$$
$$120 = 9 \cdot 13 + 3,$$
$$30 = 2 \cdot 13 + 4.$$

Thus, r_6 is the first remainder either to equal 4 or to repeat. It follows that 4/13 has a repeating decimal expansion with repetend equal to 307692, and we have $4/13 = 0.307692307692\ldots$.

Remark. It follows immediately from the repeated division algorithm procedure that equivalent fractions have the same decimal expansion. For, the quotient (place value) that arises when we divide $10a$ by b using the division algorithm is the same as the quotient (place value) when we divide $10ha$ by hb, for any positive integer h.

In Appendix A, we explain how the procedure of repeated applications of the division algorithm results in the algorithm for long division.

3. Fractions with Terminating Decimals

In this section, we investigate a fascinating and, perhaps, unexpected description of fractions with a terminating decimal expansion. In Section 1, we discussed some examples to illustrate that every fraction having a terminating decimal expansion is equivalent to a decimal fraction and, conversely, every fraction equivalent to a decimal fraction has a terminating decimal expansion. We justify the general case now.

OBSERVATION 3.1. *If a fraction a/b has a terminating decimal expansion then a/b is equivalent to a decimal fraction. Conversely, if a fraction a/b is equivalent to a decimal fraction, then a/b has a terminating decimal expansion.*

First, we assume that the fraction a/b has a terminating decimal expansion. It is just a simple matter of adding the fractions in the expansion using, as common denominator, the largest power of 10 that occurs as a denominator in the expansion. Thus, if

$$\frac{a}{b} = 0.q_1 q_2 q_3 \cdots q_k = \frac{q_1}{10} + \frac{q_2}{10^2} + \cdots + \frac{q_k}{10^k},$$

then, applying the rules of addition of fractions, we may write

$$\frac{a}{b} = \frac{q_1 10^{k-1} + \cdots + q_k 10^0}{10^k}.$$

Suppose, on the other hand, that a/b is equivalent to a decimal fraction and we have
$$\frac{a}{b} = \frac{c}{10^k},$$
for some positive integers c and k. In this case, we reverse the argument above, by writing c in base 10 :
$$c = q_1 10^{k-1} + \cdots + q_k 10^0.$$
Since $a < b$, the highest power of 10 that can occur in the base 10 expansion of c is $k - 1$. Thus, we have
$$\frac{a}{b} = \frac{q_1}{10} + \frac{q_2}{10^2} + \cdots + \frac{q_k}{10^k} = 0.q_1 q_2 q_3 \cdots q_k.$$
Be alert to the fact that for some j, where $j < k$ because $a > 0$, the first j decimal digits $q_1, q_2, \ldots, q_j$ may possibly be 0.

Since equivalent fractions have the same decimal expansion, we come to the following conclusion.

To describe fractions with terminating decimal expansions, we must identify fractions that are equivalent to decimal fractions.

Suppose the fraction a/b, in lowest terms, is equivalent to a decimal fraction, that is
$$\frac{a}{b} = \frac{c}{10^n},$$
for some positive integers c and n. The equation
$$a \cdot 10^n = bc$$
tells us that the denominator b divides $a \cdot 10^n$. We apply the very important Observation 2.2 in Seminar 4 which states, in different notation, that if a and b are relatively prime integers, and if b divides a product with a as one factor, such as the product $a \cdot 10^n$, then b must divide the other factor, which in this case is 10^n. Consequently, *if a/b is equivalent to a decimal fraction $c/10^n$, for some integer c, then b divides 10^n.* Thus, it is necessary to determine the prime divisors of powers of 10.

✳

Seminar Exercise.
  (i) Write down the prime factorizations of 10, 100, 1000 and 10000. What is the prime factorization of 10^n, where n is a positive integer?
 (ii) Find a fraction a/b, in lowest terms, that is equivalent to, but not equal to, a decimal fraction $c/10^3$, for some integer c. Find a fraction a/b, in lowest terms, that is equivalent to, but not equal to, a decimal fraction $c/10^4$, for some integer c. If n is any positive integer, find a fraction a/b, in lowest terms, that is equivalent to, but not equal to, the decimal fraction $c/10^n$, for some integer c.
(iii) For each of your examples a/b in (ii), write down the prime factorization of the denominator b. What do you observe about these denominators?

✳

Comments on the Seminar Exercise.

(i) We have

$$10 = 2 \cdot 5; \ 100 = 10^2 = 2^2 \cdot 5^2; \ 1000 = 10^3 = 2^3 \cdot 5^3; \ 10000 = 10^4 = 2^4 \cdot 5^4.$$

More generally, for any positive integer n, $10^n = 2^n \cdot 5^n$.

(ii) Here are some examples of fractions equivalent to, but not equal to, decimal fractions.

$$\frac{19}{50} = \frac{19 \cdot 2}{50 \cdot 2} = \frac{38}{100}; \ \frac{7}{40} = \frac{7 \cdot 25}{40 \cdot 25} = \frac{175}{1000}; \ \frac{9}{16} = \frac{9 \cdot 625}{16 \cdot 625} = \frac{5625}{10000}.$$

(iii) For the fractions above, the denominators factor as follows.

$$50 = 2 \cdot 5^2; \ 40 = 2^3 \cdot 5; \ 16 = 2^4;$$

Observe that each denominator is a product of powers of 2 and/or powers of 5.

✳

The exercise gathered evidence for the following observation. We remind you that $a^0 = 1$, for all positive integers a.

OBSERVATION 3.2. *If a/b is a fraction in lowest terms with $0 < a < b$, i.e., with $0 < a/b < 1$, and if*

$$\frac{a}{b} = \frac{c}{10^n},$$

for some positive integers c and n, then

$$b = 2^h 5^k, \ where \ 0 \le h \le n, \ 0 \le k \le n.$$

We verify the observation. (The argument should be familiar to you now.) Let a/b be a fraction in lowest terms with $0 < a/b < 1$. If $a/b = c/10^n$, for some positive integers c and n, then the equation

$$a \cdot 10^n = bc$$

implies that the denominator b divides $a \cdot 10^n$. Since b and a are relatively prime, by Observation 2.2 in Seminar 4, it follows that b divides 10^n. But, we know, from Seminar 5, Section 1, how to describe all positive divisors of an integer once we know its prime factorization. The positive divisors of $10^n = 2^n 5^n$ are exactly the positive integers of the form $2^h 5^k$, where $0 \le h \le n$ and $0 \le k \le n$. Consequently, the denominator b must be of this form.

The converse of the statement in Observation 3.2 is also true. Note that its verification gives a "recipe" for finding an equivalent fraction. It is also a nice exercise on the laws of exponents.

OBSERVATION 3.3. *Let a/b be a fraction in lowest terms with*

$$b = 2^h 5^k, \text{ for some integers } h \text{ and } k.$$

Then there is a positive integer c such that $a/b = c/10^n$, where n is the larger of the two values h and k, or, in the case $h = k$, $n = h = k$.

We separate the verification of the observation into three cases and also look at an example of each case.

Case 1. Suppose that $h = k$. Then, we set $n = h = k$. We have $b = 2^n \cdot 5^n = (2 \cdot 5)^n = 10^n$, and it follows that

$$\frac{a}{b} = \frac{a}{10^n}$$

is itself a decimal fraction.
For example, if $a/b = 3/(2^2 5^2)$, then $a/b = 3/(2 \cdot 5)^2 = 3/100 = 0.03$

Case 2. Suppose that $h < k$. Then, we set $n = k$ and observe that $k = h + (k - h)$. We can write

$$\frac{a}{b} = \frac{a}{2^h 5^k} = \frac{a \cdot 2^{k-h}}{2^h 5^k \cdot 2^{k-h}} = \frac{a \cdot 2^{k-h}}{10^k}.$$

Thus, a/b is equivalent to the decimal fraction with numerator $a \cdot 2^{k-h}$ and denominator 10^k.
For this case, consider the example

$$\frac{a}{b} = \frac{3}{250} = \frac{3}{2 \cdot 5^3}.$$

Then,

$$\frac{a}{b} = \frac{3 \cdot 2^2}{2 \cdot 5^3 \cdot 2^2} = \frac{12}{1000} = 0.012.$$

Case 3. Suppose that $h > k$. Then, we set $n = h$, and we observe that $h = (h - k) + k$. We can write

$$\frac{a}{b} = \frac{a}{2^h 5^k} = \frac{a \cdot 5^{h-k}}{2^h 5^k \cdot 5^{h-k}} = \frac{a \cdot 5^{h-k}}{10^h}.$$

For this case, take, for example, $a/b = 7/16 = 7/2^4$. Note that the exponent of 5 in this example is 0. We have

$$\frac{7}{2^4} = \frac{7 \cdot 5^4}{2^4 \cdot 5^4} = \frac{7 \cdot 5^4}{10^4} = \frac{4375}{10000} = 0.4375.$$

COROLLARY 3.4. *A fraction a/b, in lowest terms, with $0 < a/b < 1$, has a terminating decimal expansion precisely when $b = 2^h 5^k$, for some nonnegative integers h and k.*

Note that the inequalities $0 < a/b < 1$ imply that at least one of h and k must be nonzero. The corollary follows immediately from Observations 3.1, 3.2 and 3.3.

Summary: To write a fraction a/b with denominator a product of powers of 2 and/or powers of 5 as a decimal fraction, we multiply numerator and denominator of the given fraction by whatever power of 2 or 5 is needed to make the exponents of 2 and 5 in the denominator equal.

✳

Seminar Exercise. Tell whether each fraction below is or is not equivalent to a decimal fraction. If it is, find a decimal fraction equivalent to it. If it is not, explain why not.

$$\frac{3}{7}; \ \frac{7}{50}; \ \frac{11}{240}; \ \frac{3}{160}; \ \frac{11}{125}.$$

✳

Comments on the Seminar Exercise. The fraction $3/7$ is not equivalent to a decimal fraction because 7 is a prime number and is not divisible by 2 or 5. Remember, the primes dividing the denominator of a fraction equivalent to a decimal fraction can be only 2 and/or 5. Thus, $11/240$ is not equivalent to a decimal fraction since 240 is divisible by the prime 3.
For the fraction $7/50$ we have

$$\frac{7}{50} = \frac{7}{2 \cdot 5^2} = \frac{7 \cdot 2}{2 \cdot 5^2 \cdot 2} = \frac{14}{2^2 5^2} = \frac{14}{100} = 0.14$$

For the fraction $3/160$ we have,

$$\frac{3}{160} = \frac{3}{2^5 \cdot 5} = \frac{3 \cdot 5^4}{2^5 \cdot 5 \cdot 5^4} = \frac{1875}{2^5 5^5} = \frac{1875}{100000} = 0.01875$$

For the fraction $11/125$ we have,

$$\frac{11}{125} = \frac{11}{5^3} = \frac{11 \cdot 2^3}{5^3 \cdot 2^3} = \frac{88}{1000} = .088$$

4. Fractions with Nonterminating Decimals

We have seen many examples of fractions with nonterminating decimal expansions. In Observation 2.2, we determined that if the decimal expansion of a fraction a/b does not terminate then it is *repeating*, that is, a sequence of place values, called the *repetend*, recurs over and over, without interruption. Moreover, we discerned that the repetend starts repeating at the 10^{-b} ths place or earlier, and has length at most $b - 1$.

We introduce the following notation for nonterminating decimals. We write the repetend once and put a "bar" over it. For example, we write

$$\frac{1}{12} = 0.08333333 \cdots = 0.08\overline{3}$$

and

$$\frac{7}{11} = 0.63636363\cdots = 0.\overline{63}$$

Here are some examples of previously calculated nonterminating decimal expansions. We write them as sums of decimal fractions with their repetends as numerators.

$$\frac{5}{11} = 0.\overline{45} = \frac{45}{10^2} + \frac{45}{10^4} + \frac{45}{10^6} + \cdots$$

$$\frac{1}{7} = 0.\overline{142857} = \frac{142857}{10^6} + \frac{142857}{10^{12}} + \frac{142857}{10^{18}} + \cdots$$

$$\frac{4}{9} = 0.\overline{4} = \frac{4}{10} + \frac{4}{10^2} + \frac{4}{10^3} + \cdots$$

The fraction $5/11$ has repetend 45 of length 2. The fraction $1/7$ has repetend 142857 of length 6, the longest possible, whereas $4/9$ has repetend 4 of length 1. In these examples, the repetend begins in the tenths place. In other cases, such as

$$\frac{1}{6} = 0.1\overline{6}, \quad \frac{2}{15} = 0.1\overline{3} \quad \text{and} \quad \frac{5}{14} = 0.3\overline{571428}$$

the repetend begins in the hundredths place. For the fraction

$$\frac{1}{12} = 0.08\overline{3}$$

the repetend begins in the thousandths place.

Observe that, in the examples above, with repetend that does not start in the tenths place, each of the denominators, namely 6, 12, 14 and 15, has a factor of 2 or 5 and another prime factor $p \neq 2$ or 5. Whereas, in the examples $5/11$, $1/7$ and $4/9$ where the repetend starts in the tenths place, the denominator is relatively prime to 10. This leads us to the possibility that the prime factors of the denominator of a fraction a/b and their relation to 2 and 5, the prime factors of 10, might determine the form of the decimal expansion of the fraction. In fact, this is true. We review the possible distinct forms which the prime factorization of the denominator b may take:

(i) $b = 2^h 5^k$, for nonnegative integers h and k,

(ii) b is relatively prime to 10;

(iii) b is not relatively prime to 10 and has prime factors other than 2 or 5.

The following observation is packed with powerful, practical information on decimal expansions.

OBSERVATION 4.1. *Let a/b be a fraction, in lowest terms, with $0 < a/b < 1$. The decimal expansion of a/b is determined by the prime factors of b in the following way.*

(i) *If $b = 2^h 5^k$, for some nonnegative integers h and k, then a/b has a terminating decimal expansion.*

(ii) *If b is relatively prime to 10, then a/b has a nonterminating decimal expansion and the repetend begins in the tenths place.*

(iii) *If b is not relatively prime to 10 and has prime factors other than 2 or 5, then a/b has a nonterminating decimal expansion and the repetend begins in the 10^{-j}ths place, where $j \geq 2$.*

The case where $b = 2^h 5^k$ was explained completely in Section 3. The analysis needed to verify the remaining two cases is somewhat complicated and we will not delve into that here. However, we will indicate, mostly with examples, a method used to verify (ii) when b is a prime not equal to 2 or 5. The length of the repetend may also be determined in this case. This is fun to work out, and is a fact that surprises many teachers.

It may also surprise you that congruence plays a role in repeating decimals. Here, we will explain how congruence is used to determine when the repetend begins in the decimal expansion and its length in the case where the denominator b of a/b is a prime not equal to 2 or 5.

So, for the remainder of this seminar, we assume that the fraction under study is a/p, where p is a prime number not equal to 2 or to 5, for example, the fractions $1/3$ and $7/11$. (This is a special case of Observation 4.1 (ii).)

We are particularly interested in congruences mod p, where p is a prime number and $p \neq 2$ or 5. Recall, see Seminar 6, Section 4.1, the Powers Property. It states that if $a \equiv b \pmod{p}$, then

$$a^k \equiv b^k \pmod{p},$$

for all positive integers k. For example, for the prime $p = 3$, we have $10 \equiv 1 \pmod 3$, since $10 = 3 \cdot 3 + 1$. Consequently, $10^k \equiv 1 \pmod 3$, for every positive integer k. Here is another example. For the prime $p = 7$, we have $10 \equiv 3 \pmod 7$, since $10 = 7 + 3$. It follows that

$$10^2 \equiv 3 \cdot 3 \equiv 9 \equiv 2 \pmod 7$$

$$10^3 \equiv 10 \cdot 10^2 \equiv 3 \cdot 2 \equiv 6 \pmod 7$$

$$10^4 \equiv 10^2 \cdot 10^2 \equiv 4 \pmod 7$$

$$10^5 \equiv 10^3 \cdot 10^2 \equiv 12 \equiv 5 \pmod 7$$

$$10^6 \equiv (10^2)^3 \equiv 2^3 \equiv 1 \pmod 7$$

We stopped computing powers of 10 mod 7 at 10^6 because it is the *smallest power of 10 that is congruent to 1 modulo 7*. This has relevance to the decimal expansion of fractions $a/7$, where $1 \leq a \leq 6$, as we will explain momentarily.

✳

Seminar Exercise.

  (i) For the prime $p = 11$, find the smallest power k of 10 such that $10^k \equiv 1 \pmod{11}$.

(ii) For the prime $p = 13$, compute the powers of 10 mod 13. End your computation with the least power k of 10 such that $10^k \equiv 1$ (mod 13). Since $100 = 7 \cdot 13 + 9$, you can start your computation with

$$10 \equiv 10 \quad (\text{mod } 13)$$
$$10^2 \equiv 9 \quad (\text{mod } 13)/$$

✳

Comments on the Seminar Exercise.

(i) For the prime $p = 11$, since $100 = 9 \cdot 11 + 1$, we have

$$10 \equiv 10 \quad (\text{mod } 11)$$
$$10^2 \equiv 1 \quad (\text{mod } 11).$$

(ii) For the prime $p = 13$, we have

$$10^1 \equiv 10 \quad (\text{mod } 13)$$
$$10^2 \equiv 7 \cdot 13 + 9 \equiv 9 \quad (\text{mod } 13)$$
$$10^3 \equiv 10^2 \cdot 10 \equiv 9 \cdot 10 \equiv 12 \quad (\text{mod } 13)$$
$$10^4 \equiv (10^2)^2 \equiv 81 \equiv 6 \cdot 13 + 3 \equiv 3 \quad (\text{mod } 13)$$
$$10^5 \equiv 10 \cdot 10^4 \equiv 30 \equiv 4 \quad (\text{mod } 13)$$
$$10^6 \equiv 10 \cdot 10^5 \equiv 10 \cdot 4 \equiv 1 \quad (\text{mod } 13).$$

✳

OBSERVATION 4.2. *Let a/p be a fraction in lowest terms with $0 < a < p$, where the denominator p is a prime $\neq 2, 5$. If k is the smallest integer such that $10^k \equiv 1$ (mod p), then the decimal expansion of a/p is nonterminating with repetend that begins in the tenths place and has length k.*

We outline how the observation may be verified. Since $10^k \equiv 1$ (mod p) means that $10^k - 1$ is divisible by p, there is an integer m such that

$$a(10^k - 1) = mp.$$

Consequently, we have

$$\frac{a}{p}\left(10^k - 1\right) = m.$$

After some algebraic manipulations and substitution of a geometric series, this equation leads to a nonterminating sum of decimal fractions with the repetend m as numerator:

$$\frac{a}{p} = \frac{m}{10^k} + \frac{m}{10^{2k}} + \cdots + \frac{m}{10^{jk}} + \cdots.$$

Thus, a/p has repetend m of length k.

For example, it follows from the observation and from our calculations of the smallest integer k such that $10^k \equiv 1$ (mod p), for various primes p, that

(i) each of the fractions $a/3$, for $1 \leq a \leq 2$, has repetend of length 1;

(ii) each of the fractions $a/7$, for $1 \leq a \leq 6$, has repetend of length 6;

(iii) each of the fractions $a/11$, for $1 \leq a \leq 10$, has repetend of length 2; and

(iv) each of the fractions $a/13$, for $1 \leq a \leq 12$, has repetend of length 6.

The repetend of a/p can be computed by dividing $10^k - 1$ by p, and multiplying the result by a. For example, the fraction $7/11$ has repetend of length 2, so we calculate its repetend by dividing $10^2 - 1 = 99$ by 11 and multiplying by 7. It follows that

$$\frac{7}{11} = 0.63636363 \cdots = 0.\overline{63}.$$

We refer the interested reader who would like more details of the computation for Observation 4.2 to Chapter IX of Hardy and Wright's book *An Introduction to the Theory of Numbers* and to Chapter 3, Section 10.3 of P. Sally's book *Tools of the Trade*.

References

G. H. Hardy and E. M. Wright, *An Introduction to the Theory of Numbers*, Oxford Science Publications, Clarenden Press, Oxford, 1998.

Sally, Jr., P. J., *Tools of the Trade*, The American Mathematical Society, Providence, R. I., 2008.

Seminar 12

Order and the Number Line

Part I
Fractions, Decimals and Ordering

To put fractions in their rightful places on the number line, we must define order on the set of fractions and verify that equivalence of fractions respects order. The first part of this seminar on order in the set of fractions does exactly that. The second part examines fractions on the number line.

1. Positive Fractions and Negative Fractions

We begin the discussion of order by reflecting on the definition of order in the set of integers. First, we recall from Seminar 2 the definitions of positive and negative integers. The positive integers are the natural numbers

$$\{1, 2, 3, 4, 5, \ldots, 10, 11, \ldots, 57, \ldots, 348, \ldots.\},$$

which are closed under addition and multiplication. See Seminar 1, Sections 1 and 4. An integer a is negative precisely when its additive inverse, $-a$, is positive. To compare two integers a and b, we defined the symbol $<$, "less than," by stating that $a < b$ means $b - a$ is a positive integer. The symbol $>$, "greater than," is defined by stating that $a > b$ means $b < a$, i.e., that $a - b$ is a positive integer. Thus, an integer a is positive exactly when $0 < a$, and is negative when $a < 0$.

✳

Seminar/Classroom Discussion. Which fractions do you think should be called positive? Which negative? Why?

✳

Some Points to Consider During the Discussion.

(i) You might begin the discussion by considering whether the following fractions should be called positive or negative:

$$\frac{1}{2}, \ \frac{-1}{-3}, \ 5, \ \frac{-4}{3}, \ \frac{6}{-7}.$$

(ii) For coherence, it is essential that the definitions of positivity, negativity and order in the set of fractions extend, in some natural way, those of positivity, negativity and order in $\mathbb{Z}$.

(iii) In addition, the definitions of "positive" and "negative" must respect equivalence, that is, if a/b is positive and $c/d \sim a/b$, then c/d must also be positive, or if a/b is negative and $c/d \sim a/b$, then c/d must also be negative. For example, to decide whether $-1/-3$ is positive or negative, consider fractions equivalent to $-1/-3$.

✳

Keep in mind the points of the Seminar Discussion above as we state the definitions of positivity and negativity for fractions. A fraction a/b is *positive* if ab is a positive integer, or, in other words, if $ab > 0$. When this is the case, we write $a/b > 0$ or $0 < a/b$. Intuitively, we think of a fraction as positive when both numerator and denominator are positive. However, since we want the definition of order for fractions to respect equivalence, we say that a fraction is positive also in the case where both numerator and denominator are negative. This is correct, for it follows from Property (O3) (scaling by a positive integer) and Property (O6) (product of two negative integers) for $\mathbb{Z}$, that $ab > 0$ precisely when $a > 0$ and $b > 0$, or $a < 0$ and $b < 0$.

A fraction a/b is *negative* if ab is negative, or, in other words, if $ab < 0$. When this is the case, we write $a/b < 0$, or $0 > a/b$. It follows from the same properties of order in $\mathbb{Z}$, namely Property (O3) and Property (O6), that $ab < 0$ precisely when $a > 0$ and $b < 0$ or $a < 0$ and $b > 0$.

It follows immediately from the definition that the product of two positive fractions is positive. It is also clear that the sum of two positive fractions is positive if we assume that both fractions have positive denominators and, therefore, also have positive numerators. We may make this assumption because, as we know, every fraction is equivalent to one with positive denominator, addition respects equivalence and, as we will show shortly, order respects equivalence.

Note that a zero fraction, that is, any fraction of the form $0/n$, is neither positive nor negative. It needs special consideration. It is clear that the definitions for order in the set of fractions extend, in a natural way, those for order in $\mathbb{Z}$.

Returning to the fractions listed above, namely,

$$\frac{1}{2}, \quad \frac{-1}{-3}, \quad 5, \quad \frac{-4}{3}, \quad \frac{6}{-7}$$

we see that $1/2$, $-1/(-3)$ and 5 are positive because, respectively, the integer products $1 \cdot 2$, $(-1) \cdot (-3)$ and $5 \cdot 1$ are positive. The fractions $-4/3$ and $6/(-7)$ are negative because, respectively, the integer products $(-4) \cdot 3$ and $6 \cdot (-7)$ are negative.

We must verify that our definitions respect equivalence. Remember that every fraction in the equivalence class of $0 = 0/1$ has the form $0/n$, for some integer n. Consequently, this equivalence class contains no positive or negative fractions.

To check that the concepts of positive and negative fractions respect equivalence, let us first consider some examples. The fractions $1/3$ and $2/6$ are equivalent to the fraction $-1/(-3)$, and it is true that all of the products $(-1) \cdot (-3)$, $1 \cdot 3$ and $2 \cdot 6$ are positive. The fractions $-60/70$ and $18/(-21)$ are equivalent to the fraction $-6/7$, and it is true that all of the products $(-6) \cdot 7$, $(-60) \cdot 70$ and $18 \cdot (-21)$ are negative.

So far, it appears that if a fraction is positive (respectively, negative), then all of the fractions in its equivalence class are positive (respectively, negative). However, to know this is true for *every* pair of equivalent fractions, we take two fractions a/b and c/d in the same equivalence class and show that $ab > 0$ implies that $cd > 0$, and $ab < 0$ implies that $cd < 0$.

We demonstrate this as follows. First note that we may assume that a/b is not a zero fraction. Moreover, by transitivity of equivalence, we may assume that a/b is in lowest terms. Since $a/b \sim c/d$, we know there is a nonzero integer k, such that

$$\frac{c}{d} = \frac{ak}{bk}.$$

It follows that

$$cd = abk^2.$$

Now, k^2 is the square of a nonzero integer, so $k^2 > 0$, by the final Seminar Exercise in Seminar 2, Section 5. By Property (O3) for $\mathbb{Z}$, if ab is positive, then $cd = abk^2$ is positive, and if ab is negative, then $cd = abk^2$ is negative. Thus, if a/b is positive, then c/d is positive, and if a/b is negative, then c/d is negative. This completes our verification that the concepts of positive fraction and negative fraction respect equivalence. Take note of how the fact that "squares of nonzero integers are positive" came into play in our demonstration.

2. Comparison of Fractions

In the course of our work in these seminars, we have verified that every fraction has an equivalent fraction with positive denominator (for example, the fraction in lowest terms). Consequently, we may, and we will, assume, for the remainder of this part of the seminar, that **the denominators of all fractions are positive**, unless we explicitly say otherwise. This means, in particular, that a fraction a/b is positive precisely when a is positive, and a/b is negative, precisely when a is negative.

To compare fractions, we define a symbol $<$ and a relation between pairs of fractions as follows. For a pair of fractions a/b and c/d, we say

$$\frac{a}{b} < \frac{c}{d} \quad \text{if} \quad \frac{c}{d} - \frac{a}{b} \quad \text{is positive.}$$

In other words,

$$\frac{a}{b} < \frac{c}{d} \quad \text{if} \quad cb - ad > 0 \quad \text{or} \quad \frac{a}{b} < \frac{c}{d} \quad \text{if} \quad ad < bc.$$

We also say

$$\frac{a}{b} > \frac{c}{d} \quad \text{if} \quad \frac{c}{d} < \frac{a}{b},$$

that is,

$$\frac{a}{b} > \frac{c}{d} \quad \text{if} \quad ad > bc.$$

It is clear that these definitions for comparison of fractions extend, in a natural way, the definitions of comparison for integers.

Thus, we have the following rule.

Cross Product Comparison Rule for Fractions with Positive Denominators.
$$\frac{a}{b} < \frac{c}{d} \quad \text{provided that} \quad ad < bc.$$

Observe that these products ad and bc are the same as those obtained by cross multiplication. Consequently, to compare fractions, we compare the cross products of integers.

Let us order the fractions $7/13$, $3/5$, $5/8$ and $5/11$. At first glance, it appears that $5/11$ which is, by inspection, less than one half, might be the smallest, so let us check it against $3/5$ using the cross product comparison rule. We have

$$5 \cdot 5 = 25 \quad \text{and} \quad 11 \cdot 3 = 33,$$

so $5/11 < 3/5$. Next, let us compare the pair $7/13$ and $3/5$. We have

$$7 \cdot 5 = 35 \quad \text{and} \quad 13 \cdot 3 = 39.$$

By the cross product comparison rule, it follows that $7/13 < 3/5$. It is clear that $5/11$ is less than one half and that $7/13$ is greater than one half. Consequently, we have established the order:

$$\frac{5}{11} < \frac{7}{13} < \frac{3}{5}.$$

Let us compare $3/5$ and $5/8$.

$$3 \cdot 8 = 24 \quad \text{and} \quad 5 \cdot 5 = 25,$$

so $3/5 < 5/8$, and the order of the four given fractions is

$$\frac{5}{11} < \frac{7}{13} < \frac{3}{5} < \frac{5}{8}.$$

If the numerators of the fractions are negative, the comparison test is still valid. For example, to compare $(-5)/8$ and $(-2)/3$, we compare the products

$$(-5) \cdot 3 = -15 \quad \text{and} \quad 8 \cdot (-2) = -16.$$

Since $-15 > -16$, it follows that $(-5)/8 > (-2)/3$.

✳

Seminar/Classroom Activity. Order the following four fractions:

$$\frac{2}{3}, \frac{5}{9}, \frac{4}{7}, \frac{3}{8}.$$

✳

Comments on the Seminar/Classroom Activity. The ordering of the four fractions is: $3/8 < 5/9 < 4/7 < 2/3$.

✳

We must check that the comparison rule gives the same result for equivalent fractions. We consider an example first. We observed above that $4/7 < 2/3$. To show that we get the same order if we take fractions equivalent to $4/7$ and $2/3$, we take $8/14$ which is equivalent to $4/7$ and $6/9$ which is equivalent to $2/3$. We wish to show that $8/14 < 6/9$. We compute the cross products and find that $8 \cdot 9 = 72$, and $14 \cdot 6 = 84$. Since $72 < 84$, we have $8/14 < 6/9$ by the cross product comparison rule.

We leave the general case as an exercise for those who enjoy a challenge. The argument is similar to the one in the previous section used to show that the definitions of positive and negative respect equivalence.

✳

Seminar Exercise. Let a/b, c/d, a'/b' and c'/d' be fractions with b, d, b' and d' positive. Suppose that $a/b < c/d$, $a'/b' \sim a/b$ and $c'/d' \sim c/d$. Show that $a'/b' < c'/d'$.

✳

Comments on the Seminar Exercise. By equivalence of fractions, we have $ab' = ba'$ and $cd' = dc'$. We multiply both sides of the first equation by dd' and both sides of the second equation by bb' to obtain two equations:

$$ab'dd' = ba'dd' \text{ and } bb'cd' = bb'dc',$$

Next, we multiply the inequality $ad < bc$ by the positive integer $b'd'$ to obtain, by Property (O3) for $\mathbb{Z}$, $adb'd' < bcb'd'$. Using the two displayed equations above, we substitute $ba'dd'$ for $adb'd'$ and $bb'dc'$ for $bcb'd'$ in the inequality $adb'd' < bcb'd'$. The result is $a'd'bd < c'b'bd$. Since $bd > 0$, it follows, from Property (O3) for $\mathbb{Z}$, that $a'd' < b'c'$. Thus, we have verified that if $a/b < c/d$, then $a'/b' < c'/d'$.

✳

3. Properties of Order in the Set of Fractions

Here is a list of five properties of order for the set of fractions. As you will see, these properties are shared with the integers. Verification depends upon the existence of that same property in $\mathbb{Z}$.

(T) Trichotomy. For a fraction a/b, exactly one of the following holds:

$$\frac{a}{b} < 0, \frac{a}{b} = 0 \text{ or } \frac{a}{b} > 0.$$

(O1) Transitivity. Let a/b, c/d and e/f be fractions. If

$$\frac{a}{b} < \frac{c}{d} \text{ and } \frac{c}{d} < \frac{e}{f}, \text{ then } \frac{a}{b} < \frac{e}{f}.$$

(O2) Additive Property. Let a/b, c/d and e/f be fractions. If

$$\frac{a}{b} < \frac{c}{d} \text{ then } \frac{a}{b} + \frac{e}{f} < \frac{c}{d} + \frac{e}{f}.$$

(O3) Scaling by a Positive Fraction Property. If $a/b < c/d$ and $e/f > 0$, then

$$\frac{a}{b} \cdot \frac{e}{f} < \frac{c}{d} \cdot \frac{e}{f}.$$

(O4) Order and Additive Inverses. For any two fractions a/b and c/d,

$$\text{if } \frac{a}{b} < \frac{c}{d}, \text{ then } \frac{-c}{d} < \frac{-a}{b}.$$

Moreover, the converse is also true:

$$\text{if } \frac{-c}{d} < \frac{-a}{b}, \text{ then } \frac{a}{b} < \frac{c}{d}.$$

We state three additional properties for order in the set of fractions which are useful to have highlighted for reference.

(O5) Scaling by a Negative Fraction. Suppose that $e/f < 0$. If $a/b < c/d$, then

$$\frac{e}{f} \cdot \frac{a}{b} > \frac{e}{f} \cdot \frac{c}{d}.$$

(O6) The Product of Two Negative Fractions. If $a/b < 0$, and $c/d < 0$, then

$$\frac{a}{b} \cdot \frac{c}{d} > 0.$$

(R) Reciprocal Property. If

$$0 < \frac{a}{b} < \frac{c}{d},$$

then

$$0 < \frac{d}{c} < \frac{b}{a}.$$

✳

Seminar Exercise. Choose four of the eight properties of order listed above and provide demonstrations of them. Assume that the fractions have positive denominators. Since the definitions of order for fractions are formulated in terms of order in $\mathbb{Z}$, you will need to use the properties of order in $\mathbb{Z}$ in your demonstrations.

✳

Comments on the Seminar Exercise. We demonstrate the order properties listed above. We rely on the analogous order properties in $\mathbb{Z}$.

Trichotomy for fractions is interesting to verify. Consider the fraction a/b. Trichotomy in $\mathbb{Z}$ implies that exactly one of the following holds: $ab < 0$, $ab = 0$ or $ab > 0$. We examine each possibility. If $ab < 0$, then, by definition, the fraction a/b is negative and $a/b < 0$. Suppose that $ab = 0$. Since $b \neq 0$, it follows, by multiplicative cancellation, that $a = 0$. Consequently, by definition, the fraction $a/b = 0$. If $ab > 0$, then, by definition, the fraction a/b is positive, i.e., $a/b > 0$.

Next, we verify transitivity, Property (O1). We have fractions a/b, c/d and e/f with positive denominators. Our assumption is $a/b < c/d$ and $c/d < e/f$, and we must verify that $a/b < e/f$. By the cross product comparison rule, our assumption is $ad < bc$ and $cf < de$, and we must verify that $af < be$. By Property (O3) for $\mathbb{Z}$, we may multiply the first inequality by f and the second inequality by b, to obtain the inequalities

$$adf < bcf \quad \text{and} \quad bcf < bde.$$

By transitivity of order in $\mathbb{Z}$, it follows that $adf < bde$. Finally, applying Property (O3) for $\mathbb{Z}$ again, we have $af < be$, the required conclusion.

For additivity, Property (O2), we have that $a/b < c/d$, which means that $ad < bc$. We want to show that $a/b + e/f < c/d + e/f$, or, in other words, that $(af+be)/bf < (cf+de)/df$. Therefore, by the cross product comparison rule, we must verify that $(af + be)df < (cf + de)bf$. We aim to do just that. By Property (O3) for $\mathbb{Z}$, it is sufficient to show that $(af+be)d < (cf+de)b$. After applying the distributivity and commutativity rules for $\mathbb{Z}$, this last inequality becomes $adf + bde < bcf + bde$. Thus, to justify the additivity property for fractions, we must demonstrate that

$$adf + bde < bcf + bde.$$

Since $ad < bc$, it follows from Property (O3) for $\mathbb{Z}$ that $adf < bcf$. Then, by Property (O2) for $\mathbb{Z}$, we have $adf+bde < bcf+bde$, the desired conclusion.

For scaling by a positive fraction, Property (O3), we have $a/b < c/d$ and $e/f > 0$, and we must show that $a/b \cdot e/f < c/d \cdot e/f$. We have $ad < bc$, and $e > 0$, and $f > 0$. Thus, by (O3) for $\mathbb{Z}$, we have $adef < bcef$, which, after associativity and commutativity are applied, is the same as $aedf < bfce$.

Before we verify Property (O4), we take a moment to recall the definition of $-(a/b)$ found in the verification of Property (A5) in Seminar 10, Section 4. There, we defined $-(a/b)$ to be the equivalence class of the fraction $(-a)/b$. Moreover, we noted that the equality

$$-\frac{a}{b} = \frac{-a}{b}$$

indicates that $-(a/b)$ is being used to represent the equivalence class of $(-a)/b$.

Property (O4), on order and additive inverses, states that if $a/b < c/d$, then $(-c)/d < (-a)/b$. So, we have to show that $ad < bc$ implies that $(-c)b < (-a)d$, i.e., that $-cb < -ad$. This follows immediately from Property (O4) for order in $\mathbb{Z}$. Consequently, since order respects equivalence,

$$\frac{a}{b} < \frac{c}{d} \quad \text{implies} \quad \text{that} \quad -\frac{c}{d} < -\frac{a}{b}.$$

Next we consider Property (O5), scaling by a negative fraction. If $e/f < 0$ and if $a/b < c/d$, we must show that

$$\frac{e}{f} \cdot \frac{a}{b} > \frac{e}{f} \cdot \frac{c}{d}.$$

Note that since $f > 0$, and $e/f < 0$, we have $e < 0$ and $ef < 0$. By the cross product comparison rule, we have $ad < bc$ and we must show that $efbc < efad$. This follows immediately from Property (O5) for $\mathbb{Z}$.

For Property (O6), let $a/b < 0$ and $c/d < 0$. We must show that $ac/bd > 0$. Since $bd > 0$, this amounts to showing that $ac > 0$. But the hypotheses imply that $a < 0$ and $c < 0$, so $ac > 0$ follows by Property (O6) for $\mathbb{Z}$.

For Property (R), we note that the reciprocals b/a and d/c are positive because $0 < a/b$. By the cross product comparison rule, $a/b < c/d$, implies that $ad < bc$. Thus, $da < cb$ and by the cross product comparison rule, it follows that $d/c < b/a$.

✳

To conclude this section, we mention one property of order in $\mathbb{Z}$ that the set of fractions does **not** have. The set of fractions does **not** have the Well Ordering Property. Recall that the Well Ordering Property states that in every nonempty subset of the positive integers there is a smallest positive integer. We give an example of a nonempty set of positive fractions that has no least positive fraction.

Let n run over all of the positive integers, and set

$$\mathcal{U} = \left\{ \frac{1}{2}, \frac{1}{3}, \frac{1}{4}, \frac{1}{5}, \frac{1}{6}, \frac{1}{7}, \frac{1}{8}, \cdots \frac{1}{n}, \cdots \right\}.$$

$\mathcal{U}$ is a set of positive fractions, but $\mathcal{U}$ does not have a smallest positive fraction.

✳

Seminar Discussion. Is there a smallest positive fraction? Explain your answer.

✳

Comments on the Seminar Discussion. Let us see why there is no smallest positive fraction. Let a/b be a positive fraction.
Suppose that $a > 1$. Then, $b < ab$ implies that

$$0 < \frac{1}{b} < \frac{a}{b}.$$

If $a = 1$, then, since $b < b + 1$,

$$0 < \frac{1}{b+1} < \frac{1}{b}.$$

For a particular example in the case where $a > 1$, take the fraction $a/b = 3/10$. We have

$$0 < \frac{1}{10} < \frac{3}{10}.$$

For the case where $a = 1$, take $a/b = 1/10$. We have

$$0 < \frac{1}{11} < \frac{1}{10}.$$

✳

We end this section with an interesting little exercise.

✳

Seminar Exercise Let a and b be integers.

(i) Show that

$$\text{if } 0 < a < b, \text{ then } 0 < \frac{a}{b} < 1.$$

(ii) Show that

$$\text{if } 0 < \frac{a}{b} < 1, \text{ then } 0 < a < b.$$

When we represent fractions as points on the number line in the next part of this seminar, we will see that these fractions are precisely the ones that are represented by points on the number line that lie between 0 and 1, not including 0 and 1. The fractions between 0 and 1, including 0 and 1 are all fractions $0 \leq a/b \leq 1$. These correspond to integers a and b satisfying $0 \leq a \leq b$ and $b \neq 0$.

✳

Comments on the Seminar Exercise. For (i), the hypothesis $0 < a < b$ implies that $0/b < a/b < b/b$, by Property (O3) for order for fractions. Thus, $0 < a/b < 1$.

For (ii), we reverse the argument above. We apply Property (O3) for fractions and multiply $0 < a/b < 1$ through by b to obtain $a < b$.

✳

4. Ordering of Decimals

When we discuss the number line, we will find it useful to have a method for locating decimals. Of course, we already have order defined for decimals that represent fractions since we have defined the ordering of fractions. In this section, we explain how to translate the comparison of fractions into a method for comparing decimals. Our basic tool for comparing two fractions has been cross multiplication, whereas our basic tool for comparing two decimals will be place value. We consider the comparison of decimals that represent fractions a/b with $0 < a < b$.

Suppose a/b has decimal expansion

$$\frac{a}{b} = \frac{q_1}{10} + \frac{q_2}{10^2} + \frac{q_3}{10^3} + \cdots + \frac{q_j}{10^j} + \cdots .$$

Our discussion will be based on the division algorithm procedure for calculating the decimal expansion of a fraction. (See Seminar 11, Section 2.)

If we apply the division algorithm to $10a$ and b, the result is the equation

$$\frac{a}{b} = \frac{q_1}{10} + \frac{r_1}{10b},$$

where $0 \leq r_1 < b$. It follows that

$$\frac{q_1}{10} \leq \frac{a}{b} < \frac{q_1}{10} + \frac{1}{10} = \frac{q_1 + 1}{10},$$

where q_1 is the tenths place value in the decimal expansion of a/b.

In a similar way, following the methods used in Seminar 11, Section 2, from the second term $q_2/10^2$ we have

$$\frac{q_1}{10} + \frac{q_2}{10^2} \leq \frac{a}{b} < \frac{q_1}{10} + \frac{q_2}{10^2} + \frac{1}{10^2},$$

since

$$\frac{a}{b} - \left(\frac{q_1}{10} + \frac{q_2}{10^2}\right) = \frac{r_2}{10^2 b} < \frac{1}{10^2}.$$

The decimal digit q_2 is the hundredths place value in the decimal expansion of a/b. Continuing this procedure, we conclude that, for any positive integer j,

$$\frac{q_1}{10} + \frac{q_2}{10^2} + \frac{q_3}{10^3} + \cdots + \frac{q_j}{10^j} \leq \frac{a}{b} < \frac{q_1}{10} + \frac{q_2}{10^2} + \frac{q_3}{10^3} + \cdots + \frac{q_j}{10^j} + \frac{1}{10^j},$$

where q_1, q_2, q_3, $\ldots, q_j$ are, respectively, the tenths, hundredths, thousandths, $\ldots$, 10^{-j} ths place values in the decimal expansion of a/b. From this approximation of the fraction a/b by sums of decimal fractions, it follows immediately that to compare two decimal expansions of fractions between 0 and 1, we simply compare the place values at each place starting with the tenths place value. Consequently, if fractions a/b and c/d have decimal expansions

$$\frac{a}{b} = 0.q_1 q_2 q_3 \cdots q_j \ldots \quad \text{and} \quad \frac{c}{d} = 0.h_1 h_2 h_3 \cdots h_j \ldots$$

then $a/b < c/d$ if, at the first place where $q_j \neq h_j$, we have $q_j < h_j$.

For example, to compare the fractions $4/9$ and $5/11$, we can check the cross products 44 and 45 to see that $4/9 < 5/11$. Or, we can look at their decimals

$$\frac{4}{9} = 0.4444444 \ldots \quad \text{and} \quad \frac{5}{11} = 0.454545 \ldots$$

and compare place values. The tenths place values of both decimals are equal, but the hundreths place values are not. The fact that the hundredths place value 4 of $4/9$ is less than the hundredths place value 5 of $5/11$ means that $4/9 < 5/11$.

Consider the two decimals 0.5625 and 0.5375. It is immediate, by comparison of place values, that $0.5375 < 0.5625$. If, instead, we had been asked to compare the fractions these decimals represent, namely $9/16$ and $43/80$, respectively, applying the Cross Product Criterion to find that $43/80 < 9/16$ is not as direct.

The decimals $0.6501284365012843 \ldots$ and $0.6501283365012833 \ldots$ agree up to the ten millionths place. Since the ten millionths place values are 4 and 3 respectively, it follows that $0.\overline{65012833} < 0.\overline{65012843}$.

✳

Seminar/Classroom Activity.
(i) Give an example of two nonequivalent fractions with terminating decimal expansions that agree in the tenths and hundredths places and order them.
(ii) Give an example of two nonequivalent fractions with nonterminating decimal expansions that agree in the tenths place and order them.

✳

Comments on the Seminar Activity. For (i), it is easy to write down examples of terminating decimals that agree up to any particular place you choose, and then find the corresponding fractions. For example, let us write down decimals that agree in the first three places and then find their corresponding fractions. Consider 0.12345 and 0.12355. We have

$$0.12345 = \frac{12345}{100000} = \frac{2469}{20000} \quad \text{and} \quad 0.12355 = \frac{12355}{100000} = \frac{2471}{20000}.$$

These fractions are not equivalent because their decimals are not the same. We have $0.12345 < 0.12355$ because at the ten thousandths place, which is the first place where the two decimals have unequal value, $4 < 5$.

For (ii), consider that the decimal expansions of the fractions $1/3$ and $5/14$ agree in the tenths place and differ in the hundredths place.

$$\frac{1}{3} = 0.\overline{3}, \quad \frac{5}{14} = 0.3\overline{571428}$$

By comparing place values of the decimals, we see immediately that $1/3 < 5/14$.

Part II
Representing Fractions and Decimals on the Number Line

The number line is an effective device that assists our understanding of the properties of numbers interpreted as points on a line. In Seminar 2, we focused on the integer points on the number line and observed that the Well Ordering Property gives these points their characteristic disconnected appearance. In this part of this seminar, we direct our attention to representing fractions and decimals as points on the number line, and we observe how closely packed these points are.

5. Representing Fractions on the Number Line

For the correspondence between fractions and points on the number line to support our understanding of equivalence, we require that equivalent fractions correspond to the same point. This means that each point on the line that corresponds to a fraction, corresponds, in fact, to a whole equivalence class of fractions. Since we do not want a point on the number line to have more than one "label," we represent, as we often do, each equivalence class by the fraction in lowest terms in that class. This means that if a point on the line has the label $1/2$, for example, we must keep in front of our minds the fact that the fraction $1/2$ represents the whole equivalence class

$$\left\{ \frac{1}{2}, \frac{-1}{-2}, \frac{2}{4}, \frac{-2}{-4}, \frac{3}{6}, \frac{-3}{-6}, \frac{4}{8}, \cdots \right\}.$$

On the number line, we have two special points, one marked 0 and the other marked 1. The unit of measurement on the line is the distance between 0 and 1. We denote this unit of measurement by 1. As we know, the points representing integers are marked off to the right of 1 and to the left of 0, with the distance between adjacent points equal to 1. The points to the right of 0 correspond, from left to right, to the positive integers *in ascending order*, and the points to the left of 0 correspond, from right to left, to the negative integers *in descending order*. The properties of order established in Part I of this seminar dictate how we mark the points that represent fractions on the line.

A systematic way to depict fractions on an interval of the number line is to mark a point on this interval for each fraction in lowest terms with denominator 2, then each fraction in lowest terms with denominator 3, etc. For a particular choice of denominator, say the positive integer n, we proceed, preserving order at all times, as follows. We divide each of the segments of unit length into n equal parts, and note each place of division. We mark a point at each nth place that corresponds to a fraction in lowest terms. (A place that corresponds to a fraction with denominator n not in lowest terms, will already have been marked, for it corresponds to a fraction with smaller denominator.)

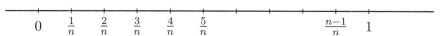

$$0 \qquad \frac{1}{n} \quad \frac{2}{n} \quad \frac{3}{n} \quad \frac{4}{n} \quad \frac{5}{n} \qquad\qquad \frac{n-1}{n} \quad 1$$

For example, consider the interval of the number line between, and including, the points 0 and 1. We denote this interval by $[0,1]$. From the Seminar Exercise at the close of Section 3, we know that it is precisely the fractions a/b, in lowest terms with $0 \le a \le b$, that satisfy $0 \le a/b \le 1$, and the properties of order mandate that the points representing these fractions lie in the interval $[0,1]$.

On the interval $[0,1]$, we use the description above to mark the points representing fractions with denominator $n = 2$, $n = 3$ and $n = 4$. As we mark these points, we observe that the ordering of fractions is preserved. For $n = 2$, there are places of division at $0/2 = 0$, $1/2$ and $2/2 = 1$. We mark a point at the $1/2$ place. For $n = 3$, there are places of division at $0/3 = 0$, $1/3$, $2/3$ and $3/3 = 1$. We mark points corresponding to $1/3$ and $2/3$.

For $n = 4$, the fractions $0/4 = 0$, $1/4$, $2/4$, $3/4$, $4/4 = 1$ are the places of division. We mark points at the places corresponding to fractions in lowest terms. So for 4, we mark points at the places of division $1/4$ and $3/4$ corresponding to the fractions $1/4$ and $3/4$. Note that there already are marked points at the other places of division for 4, namely $0/1$, $1/2$ and $1/1$, corresponding to fractions in lowest terms with denominator less than 4.

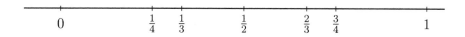

$$0 \qquad\qquad \frac{1}{4} \quad \frac{1}{3} \qquad \frac{1}{2} \qquad \frac{2}{3} \quad \frac{3}{4} \qquad\qquad 1$$

If we take $n = 5$, each of the places of division into fifths, except for $0/5$ and $5/5$, corresponds to a fraction in lowest terms, so we mark points at these places corresponding to the fractions $1/5$, $2/5$, $3/5$ and $4/5$. We add them to the picture above.

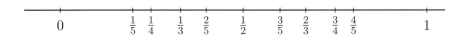

For $n = 6$, we need only mark points at the places of division $1/6$ and $5/6$ because the other fractions with denominator 6, namely $0/6 = 0/1$, $2/6 = 1/3$, $3/6 = 1/2$, $4/6 = 2/3$ and $6/6 = 1/1$, have already been represented. In general, on this same interval on the line, the fractions $0/n = 0$, $1/n$, $2/n$, $3/n$, $4/n$, ..., $(n-1)/n$, $n/n = 1$, are the places dividing the interval into $n$ths, and we mark new points on the line representing fractions in *lowest terms* with denominator n.

Here is a drawing of the portion of the number line from $-1/3$ to 1, with the points with denominators 1, 2, 3, 4, 5 and 6 marked.

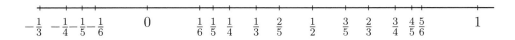

Each marked point on the number line that depicts a fraction in lowest terms is called a *rational point*. We noted earlier how isolated integer points are from one another. As we see from the depiction above, rational points are closely packed together, even though we have only marked points with denominators up to 6.

A significant feature of the use of the number line to represent fractions is the fact that order determines the position of fractions on the number line. Specifically, if a/b and c/d are fractions in lowest terms, then $a/b < c/d$ precisely when c/d appears to the right of a/b on the number line, as the following drawing illustrates. Since $-7/4 < -3/2 < 1/2 < 2/3$, we have $2/3$ lies to the right of $1/2$ which lies to the right of $-3/2$ which lies to the right of $-7/4$ on the number line.

We have set the distance between two adjacent integer points on the number line to be equal to 1. We extend the notion of distance to rational points on the line. To do this, we consider the partition of the interval between 0 and 1 into n equal parts, and we define the *length* of each part to be $1/n$. We then define the *distance between two fractions* h/n and k/n, where $0 \le h < k \le n$ to be

$$\frac{k - h}{n}.$$

The definition is extended to all fractions by stating that if $a/b < c/d$, then the *distance between a/b and c/d* is

$$\frac{c}{d} - \frac{a}{b}.$$

From the Cross Product Comparison Rule, it follows that distance is always positive.

✳

Seminar/Classroom Activity. Find all distances between the following five fractions on the number line. Draw a picture.

$$\frac{2}{5}, \ -\frac{4}{9}, \ -1, \ \frac{11}{12}, \ \frac{8}{15}.$$

✳

Comments on the Seminar/Classroom Activity.
The distance between $2/5$ and $-(4/9)$ is $2/5 - (-(4/9)) = 38/45$.
The distance between $2/5$ and -1 is $2/5 - (-1) = 7/5$.
The distance between $2/5$ and $11/12$ is $11/12 - 2/5 = 31/60$.
The distance between $2/5$ and $8/15$ is $8/15 - 2/5 = 2/15$.
The distance between $-(4/9)$ and -1 is $-(4/9) - (-1) = 5/9$.
The distance between $-(4/9)$ and $11/12$ is $11/12 - (-(4/9)) = 49/36$.
The distance between $-(4/9)$ and $8/15$ is $8/15 - (-(4/9)) = 44/45$.
The distance between -1 and $11/12$ is $11/12 - (-1) = 23/12$.
The distance between -1 and $8/15$ is $8/15 - (-1) = 23/15$.
The distance between $11/12$ and $8/15$ is $11/12 - 8/15 = 23/60$.

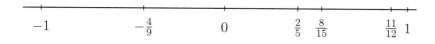

✳

Seminar/Classroom Discussion. We invite you to discuss two thought provoking questions listed below. Keep in mind particular examples of integers and fractions on the number line as you consider them.

(i) Given two integers m and n with $m < n$, is there an *integer* k such that $m < k < n$? Descriptively, we are asking if there is an integer point between any two other integer points on the number line.

(ii) If a/b and c/d are fractions in lowest terms with $a/b < c/d$, is there a fraction e/f such that $a/b < e/f < c/d$? Descriptively, we are asking if there is a rational point between any two other rational points on the line.

✳

Some Suggestions for Consideration in the Discussion.
For (i), recall that we observed how isolated and disconnected the integer points on the number line are. Certainly, for *some* integer points on this

line, such as 1 and 3, there *is* an integer point, namely 2, between, but what
about pairs of successive points such as 0 and 1 and, more generally, 4 and
5, or n and $n + 1$?

For (ii), consider how "crowded" the portion of the line looks in the drawing
of the number line from $-1/3$ to 1, with the marked points representing
fractions with denominators from 1 to 6. That another rational point might
be squeezed in between any two given rational points seems quite feasable.
Suppose a/b and c/d are two rational points on the number line. Is the
average of a/b and c/d a fraction? If so, does the point corresponding to the
average of a/b and c/d lie between a/b and c/d?

✳

Comments on the Seminar/Classroom Discussion.
Our observations in (i) about the integer points on the line are correct. For
instance, there is no integer between 0 and 1, or between any pair of successive integers of the form n, $n+1$. This fact may seem obvious, but it takes the
power of the Well Ordering Property to justify it mathematically. We will
not go into it further here. Interested readers are encouraged to give it a try.

For (ii), it *is* true that for *any* pair of fractions a/b and c/d, with $a/b <$
c/d, there is a fraction e/f such that $a/b < e/f < c/d$. Recall that the
average of a/b and c/d is the number

$$w = \frac{1}{2}\left(\frac{a}{b} + \frac{c}{d}\right).$$

We propose that the average of the fractions a/b and c/d is a fraction that
lies between a/b and c/d on the number line. The average w is a product
of two fractions, namely $1/2$ and $a/b + c/d$ and, consequently, is a fraction.
We observe, next, that it lies between a/b and c/d on the line.

OBSERVATION 5.1. *If a/b and c/d are fractions with $a/b < c/d$, then*

$$\frac{a}{b} < \frac{1}{2}\left(\frac{a}{b} + \frac{c}{d}\right) < \frac{c}{d}.$$

Descriptively, we say that the average w lies between a/b and c/d.

To verify the observation, we show first that $a/b < w$. Our hypothesis is

$$\frac{a}{b} < \frac{c}{d}.$$

Thus, by property (O2) for fractions, we have,

$$\frac{a}{b} + \frac{a}{b} < \frac{c}{d} + \frac{a}{b} \quad \text{or} \quad 2\frac{a}{b} < \frac{a}{b} + \frac{c}{d}.$$

Since $1/2 > 0$, by property (O3), the inequality

$$2\frac{a}{b} < \frac{a}{b} + \frac{c}{d}$$

implies that

$$\frac{a}{b} < w = \frac{1}{2}\left(\frac{a}{b} + \frac{c}{d}\right)$$

which is the desired conclusion. The argument to verify $w < c/d$ is analogous.

The average w of a/b and c/d is the halfway point or *midpoint* of the line segment between a/b and c/d. We leave it to you to show that the midpoint $w = 1/2(a/b+c/d)$ *is* equidistant from a/b and c/d. To do this calculate the distance between a/b and $1/2(a/b+c/d)$ and between c/d and $1/2(a/b+c/d)$ and show that these two positive fractions are equal.

✳

Seminar Discussion. How can you use midpoints to find as many rational points between a/b and c/d as you please?

✳

Comments on the Seminar Discussion. Here is one method to "populate" the line with rational points. You will undoubtedly think of many more. We have the midpoint $w = \frac{1}{2}(a/b + c/d)$ which we know lies halfway between a/b and c/d. Next, we take the average $v = \frac{1}{2}((a/b) + w)$ of a/b and w. By the argument above, it lies halfway between a/b and w. Similarly, the average $u = \frac{1}{2}((a/b) + v)$ lies halfway between a/b and v. We have three points u, v and w between a/b and c/d. For more points, we continue to take the average of a/b and the previous average. Note how these points get closer and closer to a/b from the right, but never reach a/b. It is always possible to find one more midpoint. (We could, of course, follow the same procedure for w and c/d.)

For example, if $a/b = 1/2$ and $c/d = 2/3$, then

$$w = \frac{1}{2}\left(\frac{1}{2} + \frac{2}{3}\right) = \frac{7}{12}; \quad v = \frac{13}{24}; \quad u = \frac{25}{48}.$$

On the portion of the number line drawn below, we have marked the points

$$\frac{a}{b} = \frac{24}{48} = \frac{1}{2}; \quad u = \frac{25}{48}; \quad v = \frac{26}{48} = \frac{13}{24}; w = \frac{28}{48} = \frac{7}{12}; \quad \frac{c}{d} = \frac{2}{3}.$$

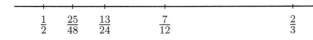

$$\frac{1}{2} \qquad \frac{25}{48} \quad \frac{13}{24} \qquad \frac{7}{12} \qquad\qquad\qquad \frac{2}{3}$$

✳

Seminar/Classroom Activity Based on the Discussion. Now that you see what fun this is, devise your own method to "populate" the portion of the number line between two rational points. This is a wonderful activity to take into a middle grade classroom when you are discussing fractions.

✳

6. Decimals on the Number Line

To place decimals on the number line, we divide the interval between two successive integers into ten equal parts of length $1/10$. We further divide each interval of length $1/10$ into ten equal parts, each of which has length $1/100$. (Of course, our accuracy in doing this is questionable after the first step.) Nonetheless, we continue to imagine repeated divisions of intervals as above. In this section, we explain how to locate the decimal representing a fraction a/b on the number line. We assume that $0 < a/b < 1$, and use the discussion of ordering decimals in Section 4.

We begin with the example $4/13 = 0.\overline{307692}$. If the interval $[0, 1]$ is divided into ten equal parts, it follows from Section 4 that

$$\frac{3}{10} < \frac{4}{13} = 0.\overline{307692} < \frac{4}{10}.$$

Consequently, we locate the point $0.\overline{307692}$ somewhere between $3/10$ and $4/10$, closer to $3/10$ than to $4/10$. If the interval between $3/10$ and $4/10$ has been divided into 10 equal parts with each of length $1/100$, then we use the approximation

$$\frac{3}{10} + \frac{0}{10^2} \leq \frac{4}{13} < \frac{4}{10} + \frac{0}{10^2} + \frac{1}{10^2},$$

to see that

$$\frac{3}{10} = \frac{30}{100} < \frac{4}{13} = 0.\overline{307692} < \frac{3}{10} + \frac{1}{100} = \frac{31}{100}.$$

Thus, we locate the point $0.\overline{307692}$ somewhere between $30/100$ and $31/100$. It is impractical to consider subdividing the interval further.

For the decimal representing any fraction a/b in the interval $[0, 1]$, it follows from the discussion in Section 4 that the decimal expansion of a fraction can be used to give some idea about where the point representing a fraction should be marked. If we subdivide $[0, 1]$ into ten equal parts, we can mark a point at $q_1/10$ as an estimate of the location of a/b. If we further subdivide the interval from $q_1/10$ to $(q_1 + 1)/10$ into ten equal parts of length $1/100$, then we can approximate the location of the decimal for a/b by estimating the position of the fraction

$$\frac{10q_1 + q_2}{100}.$$

Recall, from Section 4, that if $a/b = 0.q_1q_2q_3\ldots$ and $c/d = 0.h_1h_2h_3\ldots$, then

$$0.q_1q_2q_3\cdots < 0.h_1h_2h_3\ldots$$

precisely when there is an integer j, such that $q_i = h_i$, for $i = 1, 2, \ldots j - 1$, and $q_j < h_j$. If $a/b < c/d$, then we locate the the decimals for both fractions on the number line as above. The mark for the decimal for c/d should be to the right of the mark for a/b. In practical terms, if the place values of these decimals agree at the first two places, then the points representing them will appear as though they coincide.

Appendix A. The Mysterious Long Division Algorithm

Although there are many algorithms that apply to base 10 arithmetic, for example the algorithms for adding and multiplying multi-digit integers, one of the algorithms that is taught without any references to justify it is the long division algorithm. Since the long division algorithm is a convenient way to calculate the decimal expansion of a fraction, we explain how this algorithm arises. Even if calculators are used in the classroom for long division, it is important that teachers know the origin of the long division algorithm. Teachers must understand that the long division algorithm, when interpreted properly, is simply a systematic arrangement of the procedure, described in Seminar 11, of repeated applications of the division algorithm.

Since every fraction can be written as a mixed number, we focus on division of a by b, where $0 < a < b$. Using the long division algorithm to divide a by b is the same as using the algorithm to calculate the decimal expansion of the fraction a/b. We will pair the steps of the division algorithm procedure for calculating the decimal expansion of a/b with the steps of the long division algorithm. (It may be helpful for readers to refer back to Seminar 11, Section 2.)

We use the standard long division symbol with the dividend $a.0000\ldots$ underneath, and the divisor b to the left. The quotient will appear directly above $a.0000\ldots$ on top of the line in the division symbol. Since $a < b$, we put $0.$ directly above $a.$ in the quotient.

$$\begin{array}{r} 0. \\ \hline b\,)\,\overline{a.0000\ldots} \end{array}$$

The first step of the long division algorithm is to divide $a.0$ by b. We do this by using the division algorithm to divide the integer $10a$ by the integer b :

$$10a = bq_1 + r_1, \quad \text{where } 0 \le r_1 < b.$$

As we saw in Seminar 11, Section 2, this yields a first approximation to a/b by decimal fractions:

$$\frac{a}{b} = \frac{q_1}{10} + \frac{r_1}{10b}.$$

199

The positive integer q_1 in the division algorithm is the first place value in the decimal expansion. For long division, q_1 is written in the quotient directly above the first 0 in the dividend to the right of the decimal point.

$$\begin{array}{r} 0.q_1 \\ \hline b\,)\,a.0\,0\,0\,0\ldots \end{array}$$

Next, we multiply q_1 by b, writing the product directly under $a.0$ and subtract $q_1 b$ from $10a$ which gives us r_1, since $10a - q_1 b = r_1$.

$$\begin{array}{r} 0.q_1 \\ \hline b\,)\,a.0\,0\,0\,0\ldots \\ q_1 b \\ \hline r_1 \end{array}$$

The second step in the long division algorithm is to multiply r_1 by 10, which results in the base 10 expression $r_1 0$ ("bring down the zero") and use the division algorithm to divide $10r_1$ by b :

$$10r_1 = q_2 b + r_2 \quad \text{where } 0 \le r_2 < b$$

We know, from Step 2 in Seminar 11, Section 2, that q_2 is the value of the hundredths place in the decimal expansion of a/b, so we write q_2 to the right of q_1, in the quotient, two places to the right of the decimal point. Then, we multiply q_2 by b and write it directly under $r_1 0$. We subtract $q_2 b$ from $r_1 0 = 10r_1$ which gives us r_2, since $10r_1 - q_2 b = r_2$.

$$\begin{array}{r} 0.q_1 q_2 \\ \hline b\,)\,a.0\,0\,0\,0\ldots \\ q_1 b \\ \hline r_1 0 \\ q_2 b \\ \hline r_2 \end{array}$$

Long division continues as above with each step corresponding to the repeated division algorithm procedure in Seminar 11, Section 2. We carry out the complete procedure for a numerical example.

Example. We compute the decimal expansion of $4/13$ using the long division algorithm, and compare this calculation to the calculation using the repeated division algorithm in the Seminar Exercise in Seminar 11, Section 5.

$$\begin{array}{r} 0. \\ \hline 13\,)\,4.000000\ldots \end{array}$$

The first step is to divide $4 \cdot 10 = 40$ by 13. This is done using the division algorithm:

$$40 = 3 \cdot 13 + 1,$$

with $q_1 = 3$, and $r_1 = 1$. The quotient $q_1 = 3$, the first place value in the decimal expansion, is written directly above the first 0 to the right of the decimal point.

$$\begin{array}{r} 0.3 \\ \hline 13\,\overline{)\,4.000000\ldots} \end{array}$$

Then, for the long division algorithm, we multiply 3 by 13 and write it directly under 40. We subtract $3 \cdot 13 = 39$ from 40 which gives us 1, since $10 \cdot 4 - 3 \cdot 13 = 1$.

$$\begin{array}{r} 0.3 \\ \hline 13\,\overline{)\,4.000000\ldots} \\ 39 \\ \hline 1 \end{array}$$

The next step in the long division algorithm is to multiply 1 by 10 ("bring down the zero") and use the division algorithm to divide 10 by 13 :

$$10 = 0 \cdot 13 + 10,$$

where $q_2 = 0$ and $r_2 = 10$. We know that $q_2 = 0$ is the value of the hundredths place in the decimal expansion of $4/13$, so we write 0 to the right of 3, two places to the right of the decimal point. Then, we multiply 0 by 13 and write it directly under 10. We subtract 0 from 10 and multiply the result by 10 ("bring down the zero") to obtain 100.

$$\begin{array}{r} 0.30 \\ \hline 13\,\overline{)\,4.000000\ldots} \\ 39 \\ \hline 10 \\ 0 \\ \hline 100 \end{array}$$

For Step 3, we use the division algorithm to divide $10 \cdot 10$ by 13 :

$$100 = 7 \cdot 13 + 9,$$

where $q_3 = 7$ and $r_3 = 9$. We know that 7 is the thousandths place value and we put it three places to the right of the decimal point in the quotient, and then we multiply it by 13 and put the product 91 on the line below 100. We subtract 91 from 100 and multiply the result, $r_3 = 9$, by 10 ("bring down the zero.")

$$
\begin{array}{r}
0.307 \\
13\,\overline{)\,4.000000\ldots} \\
\underline{39} \\
10 \\
\underline{\;0} \\
100 \\
\underline{91} \\
90
\end{array}
$$

For the fourth step, we use the division algorithm to divide $9 \cdot 10$ by 13 :

$$
90 = 6 \cdot 13 + 12,
$$

with $q_4 = 6$ and $r_4 = 12$. We know that 6 is the ten thousandths place value and we put it four places to the right of the decimal point in the quotient, and then we multiply it by 13 and put the product 78 on the line below 90. We subtract 78 from 90 and multiply the result, $r_4 = 12$, by 10 ("bring down the zero.")

$$
\begin{array}{r}
0.3076 \\
13\,\overline{)\,4.000000\ldots} \\
\underline{39} \\
10 \\
\underline{\;0} \\
100 \\
\underline{91} \\
90 \\
\underline{78} \\
120
\end{array}
$$

For the fifth step, we use the division algorithm to divide $12 \cdot 10$ by 13 :

$$
120 = 9 \cdot 13 + 3.
$$

So the hundred thousandths place value is 9 and $r_5 = 3$. We put 9 five places to the right of the decimal point in the quotient, and then we multiply it by 13 and put the product on the line below 120. We subtract 117 from 120, which gives us, as we know from the division algorithm, $r_5 = 3$ and multiply

$r_5 = 3$ by 10 ("bring down the zero.")

$$
\begin{array}{r}
0.30769 \\
13\,\overline{)\,4.000000\ldots} \\
\underline{39} \\
10 \\
\underline{0} \\
100 \\
\underline{91} \\
90 \\
\underline{78} \\
120 \\
\underline{117} \\
30
\end{array}
$$

For the sixth step, we use the division algorithm to divide $3 \cdot 10$ by 13 :

$$30 = 2 \cdot 13 + 4.$$

Thus, the 10^{-6}ths place value is 2 and $r_6 = 4$. We put 2 six places to the right of the decimal point in the quotient. Observe that $r_6 = 4 = a$. This means that the place values will repeat starting with 3 and the repetend is 307692.

$$
\begin{array}{r}
0.307692\ldots \\
13\,\overline{)\,4.000000\ldots} \\
\underline{39} \\
10 \\
\underline{0} \\
100 \\
\underline{91} \\
90 \\
\underline{78} \\
120 \\
\underline{117} \\
30 \\
\underline{26} \\
4
\end{array}
$$

We close this section with the following relevant question. Why bother with long division; why not use a calculator instead? Speed is what counts these days. Unfortunately, this approach has untoward consequences. If students are taught to obtain a decimal expansion only by calculator, they may unwittingly jump to incorrect conclusions. Their calculations might be impeded by the number of decimal places their calculator displays. For example, the decimal expansion of 1/19 has repetend 052631578947368421 that has length 18. Most calculators do not read to 19 places, so that it would be difficult to obtain a full decimal expansion without using long division. Even if standard textbooks never require such huge calculations, they may well be required on the job.

⁕

Seminar/Classroom Activity. For this activity, ask half of the students to calculate the decimal expansion of 5/14 using the method of repeated applications of the division algorithm, and ask the other half to find the same decimal expansion using long division. (No calculators for this activity, please.)

⁕

Comments on the Seminar/Classroom Activity. The decimal expansion of 5/14 is $0.3571428571428\ldots$, with repetend 571428 starting in the hundredths place.

⁕

Appendix B   The Pigeonhole Principle

The pigeonhole principle describes a simple counting technique. It has numerous applications.

The Pigeonhole Principle: If there are n objects (pigeons) distributed in m boxes (pigeonholes), where $n > m$, then some box contains at least two objects.

This principle is used in all areas and levels of mathematics. The most interesting applications often arise in the middle of rather involved arguments which are not appropriate here. In Seminar 10, we apply the pigeonhole principle to deduce an intriguing fact about nonterminating decimals. In the meantime, we look at some simple examples.

Example. If I have an ordinary deck of 52 cards and draw 5 cards from the deck, then at least 2 are of the same suit.

Explanation. A deck of cards has 4 suits (pigeonholes). If I chose 5 cards (pigeons) from the deck, then $5 > 4$ means that at least 2 of the cards are of the same suit.

Example. Consider the integers 1, 2, 3, 4, 5, 6, 7, 8, 9, 10. If you select any 6 of these numbers, then two of them will add up to 11.

Explanation. Observe that 5 pairs (pigeonholes) of these integers add up to 11 : 1 and 10, 2 and 9, 3 and 8, 4 and 7, 5 and 6. If I select any 6 of the integers (pigeons) from 1 to 10, then $6 > 5$ means that two of the numbers will be one of the pairs and add up to 11.

✳

Seminar Exercise. Suppose there are 5 boxes. How many objects must be placed in the boxes to ensure that one box contains at least 3 objects. Give an example. Note that we are given no information on *how* the objects are distributed into the boxes.

✳

Comment on the Seminar Exercise. 11 objects must be placed in the boxes to make sure that one box contains at least 3 objects.

✳

Seminar Activity. Give an example of an application of the pigeonhole principle.

✳

Index

Published Titles in This Series